Life as told by a Sapiens
to a Neanderthal

Juan José Millás is considered one of the most important voices in contemporary Spanish literature, and is a prolific bestselling novelist and short-story writer. He is the winner of the Premio Nadal, the Premio Nacional, and the Premio Planeta. As a journalist, he is a multi-award-winning regular contributor to both *El Pais* and the Prensa Iberica newspaper group. His work has been translated into more than twenty languages, and includes the novels *From the Shadows* and *None Shall Sleep*.

Juan Luis Arsuaga is a professor of paleontology at the Complutense University of Madrid and the director of the Human Evolution and Behaviour Institute. He is a member of the American National Academy of Sciences and of the Musée de l'Homme of Paris, a visiting professor at University College London, and a co-director of excavations at the Sierra de Atapuerca World Heritage site. He is a regular contributor to *Nature*, *Science*, and the *American Journal of Physical Anthropology*, is the editor of the *Journal of Human Evolution*, and lectures at the universities of London, Cambridge, Berkeley, New York, Tel Aviv, and Zurich, among others. The recipient of many national and international awards, he is the author of more than a dozen works.

Life as told by a Sapiens to a Neanderthal

Juan José Millás & Juan Luis Arsuaga

translated by
Thomas Bunstead & Daniel Hahn

SCRIBE
Melbourne • London

Scribe Publications
18-20 Edward St, Brunswick, Victoria 3056, Australia
2 John St, Clerkenwell, London, WC1N 2ES, United Kingdom
3754 Pleasant Ave, Suite 100, Minneapolis, Minnesota 55409, USA

First published in Spain by Alfaguara as *La vida contada por un sapiens a un neanderthal* in 2021
Published by Scribe in 2022

Typeset in Adobe Caslon Pro by the publishers

Printed and bound in the UK by CPI Group (UK) Ltd, Croydon CR0 4YY

Scribe is committed to the sustainable use of natural resources and the use of paper products made responsibly from those resources.

978 1 957363 06 6 (US edition)
978 1 922585 04 2 (Australian edition)
978 1 914484 02 5 (UK edition)
978 1 922586 53 7 (ebook)

Catalogue records for this book are available from the National Library of Australia and the British Library.

scribepublications.com
scribepublications.com.au
scribepublications.co.uk

Contents

Visiting the grandparents

Some years ago, I went to visit the archaeological site at Atapuerca, and when I got home and was asked where I had been, I said: "Seeing the grandparents."

The experience changed my life. I came back convinced that between the supposedly remote inhabitants of that renowned prehistoric settlement and myself, there was an extraordinary physical and mental proximity.

It felt to me like a wound.

The centuries that separated us were as nothing next to the millennia that connected us. As human beings, 95 per cent of our history is actually in prehistory. We have only just landed, so to speak, in this briefest of lapses we call history. This means that writing, for example, was invented only yesterday, though it has been around for five thousand years. If I closed my eyes and reached out a hand, I could have touched the old inhabitants of Atapuerca, and they could have touched me. They were in me now, but I was already in them before.

The discovery unsettled me completely.

It wasn't just that prehistory was not confined to the past, but rather that there was a real currency to it, which moved me.

The events of that period were more relevant to me than those of my own century because they explained it better. I therefore furnished myself with a basic library on the subject and started to read. As usual, the more I learned, the broader my ignorance became. I read tirelessly because the Palaeolithic was a drug and the Neolithic was two drugs and the Neanderthals were three drugs, and I found myself on the verge of multiple drug addictions when I understood that, given my age and my intellectual limitations, I would never come to know enough to be able to write an original book on the subject, which had been my intention since the visit to Atapuerca.

What kind of book?

Who knows? At times, it was a novel; at times, an essay; at times, a hybrid of novel and essay. At times, a piece of reportage or a long poem.

I quit my plan, though not the drugs.

Meanwhile, things happened. I published a novel, for example, which I was invited to present at the Museum of Human Evolution, linked to the Atapuerca settlement, in Burgos. There I met the palaeontologist Juan Luis Arsuaga, who was the museum's scientific director and co-director of the site. Arsuaga was kind enough to give me a guided tour of the institution he ran. Some of his books had been part of my rudimentary library on prehistory and evolution, and I had read them greedily, though not always getting the full benefit of them, because the palaeontologist made few concessions in his writing. In other words, I wasn't always able to be the kind of reader who was up to Arsuaga as a writer.

As an oral storyteller, on the other hand, I found him daring, seductive, agile. I would listen literally dumbstruck because with every second or third phrase he would hit yet another nail on the head with something perfectly expressed. I dearly wanted to take possession of that speech, which in some way was also mine. I noticed, too, that in order to talk about prehistory, he alluded to the present; just as, in order to refer to the present, he talked about prehistory. In short, he erased the outrageous boundaries between these two periods that mainstream education has installed in our heads and, albeit without realising it, he reinforced my sense of closeness to our ancestors. I realised, as I listened to him, that there was a continuum between the two, a continuum in which I was emotionally trapped, but which I struggled to articulate in any rational way.

Another year went by in which I carried on reading and reading until I had succeeded, I think, in opening up some cracks in the thin pane of glass that separated me from my prehistoric ancestors.

The glass that separated me from myself.

I published another novel and arranged matters so I would once again be invited to present it at the Museum of Human Evolution. I also asked my publishers whether they could organise, if at all possible, for me to have lunch with Arsuaga.

We had lunch.

During the second course, thanks to the bravery conferred on me by three or four glasses of Ribera del Duero, I decided to get right to the point.

"Listen, Arsuaga, you're a brilliant storyteller. For ignorant people like me, you explain things better when you talk than when you write."

"That's down to teaching," he said. "It forces you to come up with all manner of tricks to stop the students from falling asleep."

"The thing is," I went on, "you and I could join forces to talk about life."

"Join forces how?"

"Like this: you take me someplace, wherever you like: to an archaeological site, to the countryside, to a maternity hospital, a morgue, wherever you like, a canary exhibition ..."

"And?"

"And you tell me what it is we're looking at, you explain it to me. I'll then make your speech mine. I'll digest it, select material from it, articulate it, and commit it to the page. I think we could build a great story about existence."

Arsuaga poured himself a glass of wine and sat saying nothing for a few moments, and we then went back to eating and talking about life: about our plans, our likes and dislikes, our frustrations ... I thought my proposal hadn't interested him and he was pretending not to have heard it.

Oh well, I thought, *I'll just have to keep at it on my own.*

But when the coffee came, he looked straight at me, smiled rather enigmatically, and struck the table with the palm of his hand: "Let's do it."

And we did.

ONE

The flowering of the adenocarpus

"This is the asphodel, the flower of the Elysian fields. If you wake one day and find yourself surrounded by asphodels, it means you're in the underworld."

I looked at the flower's white petals, which opened like a hallucination before my eyes, and I wondered, given its abundance here, whether we might not be dead already — me and this gentleman who had just spoken. He, Juan Luis Arsuaga, was a palaeontologist, and I, Juan José Millás, was the one who'd been paleontologised.

The notion of having died spurred me to follow the scientist, who was now venturing into the secret depths of low vegetation that obscured a stretch of bumpy ground on which it proved hard to keep one's footing. We climbed to the top of a small V-shaped depression with a brook at the bottom. Arsuaga proceeded easily along an almost invisible path that opened up between the flowers. I tried to put my feet where he put his, but didn't always manage it, and at one point tripped and lost my balance; I got up without a sound of complaint so as to avoid his turning around and catching me in a humiliating position.

He finally reached the very top, where he stopped and waited for me to catch up, before showing me a rocky cluster of granite that called to mind the stage set of some grand theatre. Its curtain was a transparent waterfall. The eye sees; the ear hears; the inside of the nose dampens; the skin reacts gratefully to the fine horizontal rain given off by the leaping water, which is so refreshing. All our senses were on the alert, since there were challenges for all five — all five and then some, if we only made use of them.

What had we come here for? In principle, to see the waterfall, and perhaps so the waterfall could see us, too. For a moment, beneath the magnificent sun at 5 p.m. on 14 June, I noticed the divorce from nature I have experienced over the course of my life. I registered how the senses that are supposed to perceive the background trembling of that nature, but are atrophied through lack of use, had just now reawakened to offer me a few seconds, perhaps a few tenths of a second, of vast harmony with myself and with my surroundings.

Hello, waterfall, I said without opening my lips.

Welcome, Juanjo, it replied telepathically.

Perhaps I really was dead, after all.

What I was sure of was that I couldn't recall experiencing such a combination of stimuli before: the scent of the countless plants; their chromatic variety; the resonant coolness of the curtain of water; the novelty of breathing unleaded air; the buzz prompted by the fluttering of the insects … What I recalled — I'm afraid it's unavoidable — was a perfume commercial. Even in the great beyond, we are all victims of our cultural references. However, at this moment I was not sitting on the sofa, in front of the TV; I was right *inside* the commercial, as if I'd taken acid. We found ourselves deep within a wall-less temple.

"And what is nature if not a temple?" Arsuaga would, I suppose, have said, had he opened his mouth just then.

We had gone to pay our respects to the waterfall, but also, and above all, to witness the flowering of the adenocarpus, a broom-like shrub that comes out at that time of year, its various yellows giving the landscape the unusual brilliance of a Rothko.

For a moment, the sinister side to life, its threatening aspect, fell away. Life in that moment became pure movement and I a part of it, of the movement of life. And so my ideas were at times yellow like the adenocarpus; at times, white like the asphodels; and purple, at times, like the lavender, but also green, like the grass or the ears of unripe corn dotting the landscape. And each colour offered an infinite number of modulations through which my mind moved as slowly as the shadow of a cloud over the gorse.

The flowering of the adenocarpus.

In a month, perhaps sooner, when the sun began to narrow, those yellow shades would perish with the nobility of all small things as they die.

"There's nothing quite like escaping the college," Arsuaga then said.

And he was quite right. We had indeed escaped from college, since at that time, on that 14 June, he was supposed to be at the Complutense, marking exams, I think, and I was meant to be at home, trying to write the first lines of a novel whose characters had been calling to me for months. Instead, we were at the Somosierra pass, a hundred kilometres or so from Madrid, at an altitude of some fifteen hundred metres, enjoying an unforeseen day off.

Then, as we began the climb back down, the palaeontologist told me the following: "Once, about 250 million years ago, there was a mountain range here as high as the Himalayas. It eroded over time, and all we now see are its roots. The current landscape is a very recent one, the result of everything having been given over to livestock. The plants that make up scrubland are no good for pasture.

"In Spain," he continued, barely drawing breath, "there are two principal periods: the first runs from the Neolithic to 1958, at which point the social planning by the Opus Dei technocrats comes in. Until then, the countryside was a place full of people, full of voices, life here was not a sad thing, there were children running around. It would be like walking down the street. By 1970, the countryside was empty, there was nobody left. No European country now has more than 5 per cent of its population involved in agrarian activities."

"Of course," I agreed, being careful not to trip.

"By the way, there's a book I forgot to tell you about, *Evolution Man: how I ate my father*. You have to read it."

"OK, what's it about?" I asked, as if the title didn't say it all.

"Read it for yourself. Roy Lewis is the author. Now, look at those oak trees over there. There's a birch forest nearby as well."

TWO

Everything here is Neanderthal

I met Arsuaga again a couple of weeks later. In the meantime, the thought that I might be dead came and went; but when it came, I kept it from my family and everyone else around me. I played the part of a man alive, I led a normal life and went on sending my articles to the papers I write for. Many were written as if from the great beyond, though no reader ever pointed this out to me. I must say, existence took on an uncommon light during those days; everything felt more meaningful than usual.

The palaeontologist had picked me up outside my house shortly before noon, and we were now travelling in his Nissan toward the mountains surrounding Madrid.

"I've got a surprise for you."

He was doing the driving so that I could take notes in a small exercise book, with red covers, which I had bought years earlier in a bookshop in Buenos Aires, and which I had been saving to write a brilliant poem in that seemed due to arrive at any moment, but never in fact arrived. I've now stopped expecting it to.

We were silent for a while, listening to the radio, where they were scotching a rumour about some well-known figure that had been going around.

"As a species, we love rumour," said Arsuaga, picking up on the news story, "although rumour suffers by association with gossip, when actually they're quite different things. The point of gossip is to control those in charge; when one of the people in charge does something that goes against convention, against the normal way of thinking, they become the subject of gossip. How do you think evolution managed to do away with hierarchies based purely on who was the strongest?"

"I haven't a clue," I said.

"Stones. We're the only species capable of throwing objects with precision. Prehistoric Man developed this capacity, which chimpanzees don't have. Being a good shot has been essential in our evolution. It helps in the development of both the nervous system and musculature. The reason chimpanzees can't carve objects is nothing to do with their cognitive abilities; it's that they lack the necessary *physical* coordination."

The palaeontologist turned and looked at me as if to check that I was following. I made a slight gesture toward the road to remind him that he was driving. When he turned back to the wheel, I noticed how birdlike his profile was, with the nose prominent. Some time back, I think it was on the radio, I had heard somebody say that a protruding nose is a feature specific to the human face. It's flat on the other primates. Ever since then, I've always observed that appendage on people — as well as my own in the mirror — with a certain surprise. It is, if you really look at it, a most curious addition: a protuberance in the middle of the face. Arsuaga's nose, as I was saying, lent something bird-like to his appearance. His teeth, which were something of a jumble, contributed to this effect. And then there was his hair, which was white and dishevelled like the crest of certain tropical birds.

The palaeontologist sighed, smiled nostalgically, and went on: "Historians haven't taken sufficient account of this stone-throwing ability. Hit a hyena in the head with a stone, and you kill it. Dogs run at the sight of us reaching down for a stone, because if one hits them in the mouth they end up without any teeth. The throwing of stones is of signal importance. It's no good having the greatest brute strength if everyone else in the group knows how to throw stones."

Something occurred to me. "David against Goliath," I said.

"There you have it. Politics took the place of brute strength, all thanks to stones. Gossip is our way of throwing stones. A way of damaging somebody's reputation and disqualifying them from assuming the mantle of leader."

"And rumour?"

"Rumour is a form of coercion that impedes deviation from certain norms. It's a very oppressive thing, particularly in small communities. Now look at all the broom around here — the rockrose has given way completely."

We entered the Lozoya valley, with the river of the same name flowing through it, in the Guadarrama mountains, to the north-west of the Community of Madrid.

"The Guadarrama range," he said, changing tack, "is neither the highest nor the most beautiful, but you could say it is the most high-*brow*. All the *regeneracionismo* poets and thinkers wrote about it. The *regeneracionistas* weren't a class of café writers: they were tied to nature. And they are the best that came out of twentieth-century Spanish culture. After the Civil War, the countryside and sport got a bad name. An intellectual,

11

after Franco, wouldn't be seen dead in the countryside. Now, look over there: that's Peñalara."

I looked to my right and as I did, furtively glanced at my watch. It was already time for lunch, but the palaeontologist showed no signs of heading for a restaurant. When I don't eat at the right time, the drop in my blood-sugar levels or my carbs — I'm not sure which — the drop in something in my endocrine system puts me in a bad mood, so that I find it hard to listen to what people are saying.

But at that moment, after we left a small village called Lozoya behind us, we quite literally entered paradise.

A place appeared before my eyes, a place that is not of this world.

Further proof that we had died?

The sun, which was at its peak, unleashed a huge gush of light that thrilled the senses, giving rise to a perception as if of augmented reality. I opened the car window, and when I breathed in, I breathed in light, I sweated light, the light entered my pores, reached my bones, passed through their marrow, came out my back, and continued on its way down to the centre of the earth, where it would perhaps become a dark light that, inversely, would illuminate the earth's entrails. There was nobody around: no cars, no motorbikes, no bicycles. From time to time, a shadow in the shape of a bird tore through the silent matter of which the air was made.

"Are we in the Secret Valley?" I asked.

"Yes," he said, "the valley of the Neanderthals. 'Secret' because of how isolated it is."

He had talked to me about it the previous time we'd met, promising that one day he'd take me to see it. To me, this meant a visit to the grandparents', because I am a Neanderthal myself.

I've known this since school because the Sapiens kids — real bastards, the lot of them — always used to give me strange looks. It took a heroic effort to hide my Neanderthalness, and I used to spend my whole life observing them so as to imitate their behaviour, leaving me no time to devote to my studies. I put everything on hold, which only served to make me, if anything, even more Neanderthal. To look at, you wouldn't think my family was Neanderthal, which led me to believe I was adopted — an adopted idiot, of course — until I stumbled on a TV program about Neanderthals and recognised myself in the main character, who looked like a copy of me (or I was of him). My parents didn't notice a thing. Dad, who was a Sapiens through and through, said it was just as well humankind had moved on from all that.

"Why?" I'd asked.

"Because the Neanderthals," he said, "lacked symbolic capacity."

I didn't dare ask what symbolic capacity was, but I consulted the encyclopaedia and learned what a symbol was. Flags, for example. I thought they were pretty poor symbols, but I pretended to take an interest in order to pass as Sapiens. We were surrounded by symbols. My mother's Majorica pearl necklace, to give you an example, was another symbol (a status symbol). I likewise established that the Neanderthals and Sapiens had exchanged all kinds of materials, genetic material included. At first, Sapiens gave Neanderthals crystal necklaces in exchange for food, because Sapiens liked gastronomy while Neanderthals were fascinated by shiny things. Lacking in symbolic capacity as they were, I thought, they were unaware of the meaning of that sparkle, and yet were still dazzled by it. The fact was, from so much exchanging of objects, and since regular contact breeds affection, the Neanderthals and the Sapiens wound up in bed together. The Sapiens, being the smart ones, did it out of vice, while the Neanderthals, who were more naïve, did it out of love. And that's where the genetic exchange began.

My being a Neanderthal meant I had a very tough adolescence, since I didn't want girls for their money (the lack of symbolic capacity prevented my appreciating the value of banknotes), but for their sparkle. But they liked boys who had symbolic capacity — that is, the ones who understood the meaning of owning a Renault. There was no chance of exchanging genetic material with any of them. They would accept my invitations to tea, but when I offered them a helping of my semen, they'd run a mile.

It was hard; it still is. I still go around pretending I understand Sapiens, that I'm one of them, but the truth is that I suffer like a dog because they have taken their intellectual capacities to extremes that have become hard to imitate.

The palaeontologist, in short, had brought me home. That was the surprise, I suppose, that he had been referring to when we'd set off.

The sight took your breath away. It was like a platonic valley, an archetypical, hyperreal valley.

It looked like THE VALLEY.

"Beggars belief, doesn't it?" he murmured, shutting off the engine.

We got out of the car, neither of us saying anything. The palaeontologist had brought an umbrella, which he opened to shield himself from the sun, and he started up a shallow incline in search of a place from which to view the land.

"Look," he said, showing me a plant, "this is burdock. People once used it to fish with. They would drop it in a deep pool, like the one formed by the river down there, and the fish would float to the surface half-dead. And look at the briar. And the poppies. Poppies. The poppy is my flower. That red … it's beyond explanation. Don't forget to take a look at the *jarilla* flowers either."

As he named the plants, he caressed them gently with the fingertips of his left hand, still holding the umbrella in his right. As for me, I had previously only appreciated an undifferentiated mass of vegetation; now, in addition to burdock, briar, and poppies, I saw snapdragons and honeysuckle and pale flax, from which I deduced that the word, as I had long suspected, is an organ of sight. And my vision, in this case, was amplified, because wherever I looked I discovered an uncommon splendour. A simple bee, with its head buried in the depths of a flower, became an extraordinary biological exhibition.

"We in the West understand nothing," I heard Arsuaga say, more to himself than to me.

The man with the umbrella was climbing birdlike toward a bare patch of rock that protruded like the top of a poorly buried skull. It put me in mind of a sea of stone.

"Limestone," he said, reading my thoughts. "That's why there are so many caves around here. Limestone."

"What altitude are we at?"

"One hundred thousand metres. This is a tectonic valley, not a fluvial one."

"What's the difference?"

"Rivers mould to the course of a tectonic valley, because it was orogeny and tectonics — rather than the river itself — that created it. This whole mountain range, the *Sistema Central*, drains down into the Tagus and the Douro. They are what's called transversal valleys, carved out by the two rivers, which descend to their respective plateaus. That's how the fluvial network is formed. We say this valley is invisible because it can't be seen from anywhere in the sierra. That over there's the Malangosto pass; the Archpriest of Hita used to walk that way in medieval times, Sotasalbos being his parish. That's where the old poem comes from about the bear you have to lie with in order to carry on past. That was the toll. There used to be bears here."

We moved across the sea of stone, over the lids of the skulls, in direct sunlight, Arsuaga protected by his umbrella. Each pit contained a prehistoric site.

"Here," he said, "there's been great biodiversity because you've got water and you've got various layers of vegetation. Look carefully: there are ash trees by the river; there are the oaks; then you've got the pine trees; and above that, a layer of alpine brush. And at the top, the alpine grass. Going up though a defile like this is akin to traveling toward one of the Poles. This is called an Arctic-Alpine disjunction."

We reached the prehistoric site, the dig areas covered in huge plastic sheets that looked like shrouds.

"It isn't the time of year yet for excavating," said Arsuaga. "That's why everything's covered over."

I asked him whether we could lift up the plastic and go into one of the caves whose interiors could just be glimpsed underneath it, to which the answer was a reproachful no.

"There was so much going on in these caves," he went on. "We've found the remains of lions in the settlements. The lion was at the top of the food chain, meaning that where you had lions, you had bison, horses, deer, aurochs, boar … you name it. Everything. Humans did well here, because the animals had no way out. You could corral them. The worst thing for hunters was the open steppe, unless you knew how to mount a horse. In those days, Castilla was the Gobi desert."

"And how do the Neanderthals figure?"

"Everything here is Neanderthal. Look, a cave without a roof, although it had one once. This is going back fifty thousand years. We've found the teeth of a Neanderthal girl here, and the skulls of horned animals that were really trophies; the way they were conserved is not like everyday objects, but like those used in ritual."

"Symbolic behaviour?"

"There's no other way to explain it."

I wonder: so where the hell did my dad get the idea that Neanderthals lacked symbolic capacities? I became a writer so as to pretend I had them, and it turned out I really did have them all along.

In a burst of excitement, I was about to tell the palaeontologist about my Neanderthalness, but I restrained myself because we'd only met a couple of times and I didn't want to spook him this early on.

Just then, we stopped beside some rocks that looked like they had emerged from a landslide. He explained: "The rock ledge above served as a shade, a cornice, and it made for an awning like a bus shelter. As you can see, that collapsed, and these rocks are what was left. Beneath it, right where we're standing, was a Neanderthal settlement. Seventy thousand years ago, we're talking about. They made fire here, they consumed their food. They would strip their kills to the very last calorie. A bison would be reduced to a pile of bones. Stone knapping also took place, using a fairly complex method known as the Levallois technique."

While he gave a very precise description of the method,

to which, self-defensively, I did not pay too much attention, I looked around, and for a moment I could see the Neanderthal settlement in all its detail. I would have seen it even if my eyes had been closed, because the scene occurred at the same time both inside and outside my head. The first thing I noticed was that under that ledge that serves as their shelter there is no Monday or Tuesday or Wednesday, not even Sunday afternoons — how wonderful! There is no January or February or March, nor any Christmases, of course. Nor is it midday or three in the afternoon, because hours have not been invented yet; they've got plenty to occupy them in making fire, tanning the hides that protect them from the cold, and preparing tools for the hunt.

There is a group of men and women of all ages. Old ones, young ones, babies, middle-aged people. Influenced by my reading of a book by Arsuaga himself, I focus on an adolescent Neanderthal girl who is trying to extract the marrow from the bone of a herbivore. She puts the bone down on a flat stone, which she uses as an anvil, and she strikes it with a round one. At first, bone and stone slip, but after a few attempts the femur (if it is a femur) of the bison (if it is a bison) shatters, and the girl gains access to its marrow, which contains a fierce calorie hit.

The palaeontologist's voice pulled me out of my reverie: "A good deal of hunting took place here, but there wasn't any flint around to fashion weapons, so they made do with what they had, which was quartz. Quartz really isn't much good for anything, but they did a great job with it, using the knapping technique I was just telling you about."

"Right," I agreed exaggeratedly, to cover for the fact I hadn't been listening.

"And now," he said, "we'll head to the Cotos pass, get some beans from the farm, and go for fried eggs at my friend Rafa's restaurant. Then we'll head down the other side of the sierra, completing the circuit."

I'd forgotten my hunger, but when he spoke of beans I could see them, too, in my head, as well as the eggs, to which of my own accord I added some fries.

As we made our way down toward the car, I asked him when I'd be able to go into one of the sites.

"What you still haven't realised," he said, with his African umbrella over his head, "is that prehistory does not exist in the settlements — that's what ignorant people believe. Prehistory hasn't gone away; look about you, it's here, it's all around. You and I carry it inside us. The only thing the settlements contain is bones. Prehistory lies in the animal that passes by like a shadow."

The beer is cold, and the beans just right.

"What is it that defines a species?" I asked.

"First, ask yourself why there are species."

"Why are there species?"

"There are species because you say so. In nature, everything flows; nothing's static."

"But there must be some scientific consensus, I'm guessing, about what we call a species."

"If you insist: we call that which is recognised as distinct, not hybrid, a species. But then, in nature, you do get coyotes and jackals mating."

"Is *Homo neanderthalensis* a different species from *Homo sapiens*?"

"Well, that's down to you. So, are these beans good or what?"

"How am I meant to decide?"

"When does a town become a city? When does a hill become a mountain? What's the difference between a small wave and a big wave?"

"OK, but is the Neanderthal a species or not? What's your view?"

"If you insist: I say that it is. Let's get another beer."

"And yet, it hybridised with *Homo sapiens*."

"Spanish is not Arabic, but the language is littered with Arabic words. 'Almohada' for 'pillow'; that's a loan word. Genetic loans are like linguistic loans. Hybridisation is not the same as a loan."

"Right."

"Really. Nature is not made for human categories. There were animals before zoologists came along, little though a zoologist would like to accept it. We spend our lives categorising. Ah, here come the eggs. They're delicious here, you'll see."

The palaeontologist leant back in a gesture that sought to encompass the landscape, since we had found ourselves outside, on the terrace of his friend Rafa's restaurant, in the shade of a pine tree.

"Isn't this the life?" he said, with a wicked smile.

THREE

Lucy in the sky

With the arrival of summer, the palaeontologist went off on his excavations and I withdrew to my writing, fearing, of course, that such a long separation might become permanent. Arsuaga is not much of an emailer, nor a very regular user of the phone — let alone WhatsApp, of course. Arsuaga is distant, meaning there was a chance the summer might constitute a breach that would prove hard to repair in the autumn. To my surprise, I received an email on 1 August in which he set me some tasks: I should look at the footprints left on the beach by three- or four-year-old children.

"If you do," he promised me, "I'll explain bipedal locomotion to you."

He attached an image of his daughter's footprint, adding that Lucy was the height of a three- or four-year-old.

Oh my God — Lucy!

Lucy, whose remains were discovered in Ethiopia in 1974, lived about three million years ago. She was a little over a metre tall, weighed less than thirty kilos, and died at the age of around twenty. Her bones had surfaced while her discoverers were listening to "Lucy in the Sky with Diamonds".

Lucy belonged to a genus of hominid (Australopithecus) that lived in Africa until a couple of million years ago. In my fantasy, she was the first biped woman in history, and I had always felt a boundless pity for her. I imagined her coming down from a tree, standing on her hind limbs and crossing the boundary that separated the jungle from the savannah, with no weapons other than those two stunted hands at the end of her arms like two prostheses she did not yet know how to use. I've always been deeply moved by the curiosity and helplessness of this ancestor, who was so diminutive, so fragile, who had just left the treetops to conquer the surface of the earth, a place inhabited by terrible predators such as lions, but also by infectious micro-organisms for which her immune system was not yet prepared.

The reference to Lucy almost made me cry, and accordingly I did go down to the beach every day in order to look at the footprints of three- or four-year-old children and to take notes and photos. And in each of the prints I saw Lucy represented. And I thought about what a very complex architecture a foot was, much more than the showiest vaults in a Gothic cathedral. And I wondered whether at each moment in history when we'd risen another centimetre above the ground, whether we had been occupied by one more centimetre of SELF. With how many centimetres of SELF had Lucy faced the savannah?

What a strange thing, I thought, *bipedalism and the self.*

I replied to Arsuaga's message with these sentimental (even positively soppy) considerations, to which he devoted a polite line and explained how it is that we go about walking.

"The first part of the foot to strike the ground," he said, "is the heel, which constitutes the rear column of the foot arch. The weight is then transferred along the outer edge of the foot until it comes to rest on the front column of the foot arch. Next, the toes bend, and the weight of the foot is transferred to them. The final push comes from the big toe, and the leg is propelled forward like a pendulum. The footprints of the bipedal Australopithecus from 3.5 million years ago are exactly the same as those made by our children on sandy beaches. These biomechanics are now second nature to us."

I read his message on my phone, first thing in the morning, as I walked along Aguilar beach, in Muros de Nalón, Asturias. I became aware of the arched shape of my feet and took control of their rear and front columns, and I confirmed that I did indeed first plant my heel and that the energy produced by that impact was transferred to the front column through the instep, and that the force then reached my toes, the big toe in particular, which acted like a spring to propel the leg forward. Bipedalism seemed like a grammatical miracle, since this whole movement that went from the back of the foot to the front could be analysed syntactically, like a sentence. Subject, verb, direct object. I thought I would never just stroll about mindlessly again.

Later, at home, I searched on my computer for "Lucy in the Sky with Diamonds", and I listened to it over and over as I walked back and forth across the room. "Picture yourself on a boat on a river ..."

Mind-blowing.

FOUR

The fat and the muscle

In September, the palaeontologist set a date for us to meet at eight o'clock one morning at the St Jerome entrance of the Prado Museum. I was relieved to hear from him, as we hadn't exchanged any messages since the ones about Lucy, but I answered that the art gallery (that was the term that came out, art gallery) didn't open till ten. He told me not to worry; he was dealing with that.

"Be there at eight."

Those influential Sapiens, I thought, *they do one another a lot of favours.*

The day before our meeting I started to get some pain in a molar that had been bothering me for months. I called the dentist, who told me he had a free slot at exactly the time Arsuaga had set for our meeting. I turned down the dentist for fear of losing the palaeontologist, and after an absolute dog of a night, before leaving the house, I drank a vial of Nolotil that I kept in my secret potions drawer for pain and distress (but also for suicide). Those vials are actually to be used for injection, though in extreme cases can be taken orally. It's best to keep their contents in your mouth for a bit, as that allows them to filter through the mucous membranes and reach the source of the pain in seconds.

The morning air, which was cool now because of the time of year, perked me up, and I headed down my street toward the Metro station, sensing that the gum was being numbed by the drug. I was going to be fine, I promised myself in the Metro carriage, while I looked over the latest notes I had made in the red notebook I'd bought in Buenos Aires to write a brilliant poem in.

I arrived at our appointed place half an hour early, as is my custom, and I went for a wander, alert to the journey that the metamizole molecules from the Nolotil were taking along the furrows of my brain, where the optimism endorphin, if such a thing exists, had just awoken. Well-being was winning the battle against disquiet. I fell into a reflective state, under whose influence I was about to go into the Church of St Jerome to talk to myself for a bit ("He who talks to himself hopes to talk to God one day"), but I didn't, fearing I might be spotted by a literary critic who would out me on Instagram. Or Twitter.

At eight on the dot, I was at the appointed place, from where I saw the palaeontologist arriving, and a lady with him. We approached and shook hands, and he did the introductions: "Lourdes, my wife. This is Juanjo."

I greeted Lourdes politely, but I was not pleased he'd brought her. This wasn't a thing for married couples. Ever since our first meetings, Arsuaga and I had built a heterosexual male relationship that worked well. So why would we want to change it? It felt somehow as though the palaeontologist had broken an implicit agreement that might perhaps only have existed in my head. Lourdes' presence did, in short, prompt some pre-emptive jealousy in me, along with a loss of self-esteem. At some point in our visit to the Prado, I thought, he would abandon me to devote himself to her.

The door to the art gallery (that term again) was opened to us by Víctor Cageao, the institution's architect, who would also be accompanying us throughout our tour and with whom Arsuaga started to spend the whole time exchanging opinions about the latest changes to the museum. Taking the palaeontologist aside, I asked him not to give the architect quite so much encouragement, since this was diverting us from our purpose (though I didn't yet know what that was). He looked at me as if I had no manners and said: "Come on, man. He's done us the favour of letting us in at this hour of the morning."

I decided to resign myself to the situation, and joined the group seemingly naturally. Our steps as shoe-wearing bipeds echoed in the empty galleries. I couldn't get BIPEDALISM out of my head. Nor the SELF. And here we were, four BIPEDAL SELVES in search of knowledge.

Suddenly, we arrived at a circular room known as the Hall of the Muses, because it housed the nine of them, each one on a marble pillar: Calliope, Clio, Erato, Thalia … The palaeontologist explained that these statues had been a part of Hadrian's villa.

"This Hadrian," he added, "is the one from *The Memoirs of Hadrian*, which Cortázar translated."

He was referring to the novel by Marguerite Yourcenar, which the Argentine author had rendered into Spanish.

"But these muses," he concluded, before we began stopping in front of every single one, "all have clothes on, meaning they're of no use for our purposes today."

As I have said, I was unaware of our purposes, but I didn't ask in case I put my foot in it.

We made our way through further rooms, our footsteps echoing around us.

Lourdes and I stayed quiet, while Arsuaga kept up a lively conversation with the architect. The molecules of the active ingredient in the Nolotil continued, for their part, to explore all the ridges inside my skull, provoking bursts of optimism, though also reducing me to mutism, as if their passage through the language area had restricted its capabilities. It was a resentful silence in part, owing to the presence of Lourdes and the architect, and in part a silence imposed by the numbing of the gum, as I feared I might rouse the nerve in the molar if I opened my mouth.

En route to our destination, whatever that was, we stopped in front of a bronze head, the sight of which could leave nobody indifferent, or at least nobody who had taken a strong analgesic two hours earlier.

"They now think this is the head of Demetrius I of Macedonia, also known as Poliorcetes," said Arsuaga. "But I'd like to believe it's Alexander the Great's."

I approached it, the head, which was level with my own, and I understood that it must originally have belonged to a statue of huge dimensions. A very recent restoration had returned the bronze to its original colour. Despite some thin cracks and a bit of light wear to the nose, it was perfectly possible to appreciate its features, which are those of a very beautiful young man, very athletic, with curly hair that leaves his admirable ears uncovered, and a pair of thick, sumptuous, slightly parted lips. The whole ensemble conveys a degree of narcissistic serenity that even a latest-generation sedative could not offer. From the young man's expression one can deduce a Platonic sort of self-possession. In spite of the mental complexity that can be divined behind the heavy brow, he seems to be troubled by no concerns whatsoever. His eye sockets, which are empty, nonetheless look at his viewer as if those hollows contained invisible pupils.

"Look at him in profile," I heard Arsuaga say.

The palaeontologist possesses a diabolical ability for finding just the right place to put the camera when observing the ancient world. I looked at the head in profile, and from this perspective, too, it proved to be of an unbridled perfection. I would gladly have taken this head from around 300 B.C.E. home with me.

"Looks like he's just shaved," I said, as I examined the smooth surface of his cheeks.

"Excellent observation," said Arsuaga, renewing my self-esteem. "The first great historical personage to shave off his beard was Alexander the Great. Whereas his father, Philip II, did not. Look it up on Google and you'll see. Take note: the beard is a secondary sexual characteristic."

"Alexander was a homosexual?"

"Alexander is unclassifiable, inaccessible, he's like a god. There is no way to get purchase on him; he can be neither studied nor categorised. His contemporaries wondered where he had come from. He himself asked the oracle who his father was, and the oracle said it was Zeus. At any rate, the question of beards is apt, because our subject today is the differentiation of the sexes."

So, I said to myself, *we're doing the differentiation of the sexes now*. We had come to the museum on matters venereal. The palaeontologist had seemed to be wandering aimlessly, but he never failed to have an object in mind.

Very regretfully, we left behind the bronze head of

Alexander (or Poliorcetes, the King of Macedonia, depending on whom you ask), our steps taking us to a hall in which we were met by the marble sculpture of a naked youth.

"Here we have one of the best replicas in existence of the Diadumenos sculpted by Polyclitus in the fifth century B.C.E.," said Arsuaga. "An unreserved exaltation of youth. The Diadumenos," he continued, "is that young athlete who's tying a strip of cloth around his head. When they found this one, they failed to identify it correctly; it was missing a right arm, and they put it down as being an archer. It should have its arms all the way up, hands level with the forehead. But anyway, there you have it, in all its physical glory. The Greeks invented the human body, because the human body isn't like this. This knee muscle, for example, doesn't exist. But the whole ensemble is splendid, added to which it's got one of the best arses in the history of art."

We stood around the statue and looked at the nude from head to foot, fascinated by the fake transparencies offered by the marble and by the rhythm of the stone, as the viewer's gaze slides from the athlete's head to the torso, and from the torso to the waist, and from the waist to the thighs, as if passing through the lines of a sonnet. The position of his feet reminded me of the subject of BIPEDALISM, but Arsuaga insisted that today was about sexual dimorphism, so we looked closely at the young man's body, paying particular attention to his secondary characteristics.

Later, as we were walking away from the statue, the palaeontologist explained that Darwin discovered two principles: "Darwin saw himself like Newton with his law of universal gravitation. He tried to find laws that would elucidate dimorphism. Natural selection explained biological adaptation, and the way every animal has its corresponding niche in nature. And that solved almost everything for him, but there had to be another law as well: that of the selection of the mate. In other words, adaptation to one's environment is not the whole picture. There's also the struggle to reproduce, and out of that you get some real curiosities: the tail of a peacock, for instance, is not only adaptive, it's also a real hindrance to them."

For a moment, it occurred to me that the palaeontologist's wisdom was his peacock tail. But my fear was that he wouldn't show it in my presence, and I suddenly understood why he had brought his wife along. May God forgive me for this heterosexual attack of jealousy, which I tried to banish from my head so as not to lose sight of the subject of sexual differentiation.

"In mammals," I heard him say, as I stopped daydreaming, "the male is the strongest. Whereas the opposite is the case with birds. But size and strength aren't everything. The grouse, to take an obvious example, sings. There are primary sexual organs, and there are secondary sexual organs. The former have to do with reproduction; the latter, with choosing a mate."

"And apart from the beard, what are the secondary sexual characteristics for us?" I asked.

"Everything and anything that serves to distinguish a man from a woman is, seen from any point of view, a secondary characteristic. For example, women have large breasts, which is not the case in other primate species. The female chimpanzee does not have such a thing. Imagine a chimp with nice boobs and a nice waistline."

I did imagine it, and smiled.

"The beard," Arsuaga added, "has no biological function. It's there purely to catch the eye."

"Right," I said, somewhat out of it; perhaps the Nolotil had reached its moment of greatest effectiveness, because I was struggling slightly to focus. Besides, I hadn't yet recomposed myself following the emotional impact produced by the sight of the bronze of Alexander and the nude Diadumenos.

After a brief mental absence, which could only be explained by the secondary effects of the drug, I realised we had moved forward and stopped in the middle of one of those corridors that connects each room with the next. Arsuaga was looking my way as if waiting for me to say something, but I didn't know what to say because I didn't know what he'd asked me. So he stepped in.

"If you want to know why it is that you like women," he said, "you'll have to ask yourself what it is that women have in common."

"I don't know where you're going with this," I replied, trying to orientate myself spatially.

"What I've been trying to get at is that, over the course of the history of art, the image of women has changed more than that of men has. That's my opinion."

"Right," I said.

"Norms around the feminine," he continued, "are more variable, but underneath this diversity there has to be something immutable. It isn't possible for everything to be culture; biology has to be somewhere in the mix as well. Do you see?"

"More or less."

"When it comes to such questions, there's very little compromising: for some, everything is culture, and for others, biology. Culture is another layer. How can I put it: we have eyes, and we also have microscopes, which allow us to access those places the eye cannot reach. Eyes are biology, and the microscope is culture. OK?"

"OK," I agreed, satisfied that the palaeontologist had finally focused on me while his wife and the architect were chatting a few steps ahead.

"Fine, so what is it that makes a man attractive to a woman, and a woman to a man?"

"If you ask me, I don't know; if you don't ask me, I do," I replied, parodying Saint Augustine's answer to a question about time.

"The possibility of reproducing," he said, ignoring my joke. "Sexual attractiveness has a lot to do with fertility. You choose them, and they choose you, above and beyond passing cultural norms, based on this question of a biological order."

From that moment, the gum unnumbed — if the verb *to unnumb* existed — and a murderous throbbing, coming from the nerve in the molar, began to send Morse signals to my brain. I started to panic.

"Are you okay?" the palaeontologist asked.

"Fine," I said.

"Let's go and take a look at Rembrandt's *Judith*, and Rubens's *The Three Graces*."

On our way to those paintings, we stopped at Dürer's *Adam* and *Eve*.

"Look how modern they are," exclaimed Arsuaga. "They still haven't eaten the apple."

"And the apple was sex?" I asked.

"Take note," he said, ignoring my question: "the fat and the muscle."

"The fat and the muscle," I repeated. "OK."

I pretended to be looking at Rembrandt's *Judith* and Rubens's *The Three Graces*, but the truth was, I was blind with pain. I should have brought along another vial of Nolotil for the road. Then I suffered another absence, after which I found myself in front of Goya's *Majas*. I hadn't the faintest idea how much time had passed between the Dutch painter and the Spaniard.

"Here you have the solution to the enigma," I heard the palaeontologist enthuse. "Fat and muscle. Look at the proportions of the waist and hips in *The Nude Maja*."

I looked.

"Those proportions transmit a sense of fertility, and they have remained constant in representations of women from prehistoric times till the present day. This woman is a fertile woman. She ovulates. Everything else can change according to fashion, but not this. In men, muscle predominates; in women, fat. The amounts of fat or muscle may change, but not their distribution. Women's curves, which are so attractive to us men, are down to this distribution. Don't you find it astonishing?"

"What?"

"The sexual dimorphism."

"Yes," I said.

"Every species has things that set it apart sexually. I'm explaining ours to you. And we've skipped modern art entirely. Think of Modigliani's women."

Oh, God, Modigliani's women, I said to myself with a wail of pain.

FIVE

The revolution of the small

The palaeontologist sent me an email arranging to meet at nine on a mid-November morning at the entrance to Madrid's Chamartín market. He said he was coming from Dublin and going on to Burgos, but that he had three hours free to show me something.

I saw him spring out of a car like a teenager. He looked happy and spirited. Dublin had done him a world of good. When Arsuaga is happy, he communicates very well, and supplements his eloquence with a sense of humour that is full of compassion for humanity and all its foibles.

Following the requisite greetings, we went into the market and immediately stopped at a fruit and veg stall that looked like a whole syntax of shapes and colours. There, ordered with great care, you could find fruit, vegetables, legumes, tubers. The chromatic contrasts of the different types competed with those of all the flags in the world: reds, yellows, blues, browns, violets, oranges, greens …

"Although we're going to be talking about the Palaeolithic," said Arsuaga, "everything here is Neolithic inasmuch as everything is cultivated."

"So a cultivated lettuce could be the flag of the Neolithic?" I said in what I thought was a rather good witticism, but which was received with indifference.

"Let's stay on track," suggested the professor.

As I was saying, we were standing at a large vegetable stall on one corner, at which five or six people were working who soon began to find our presence strange, because we were standing facing one another — me holding a tape recorder just a few centimetres from the palaeontologist's mouth. I had put it so close for fear that the market's ambient noise would prevent his words from being faithfully captured. We were two outlandish figures who were bothering the customers, but I was the only one who seemed to notice this, because the professor was continuing blithely on, oblivious to the bafflement we were causing around us.

As the minutes passed, the racket grew and the shouts intensified, so that I needed to get even closer to the palaeontologist to be able to hear him properly. From one end of the vegetable stall to the other, the sellers called out to their colleagues for a kilo of onions or a bunch of leeks. The constant ringing of tills gave a sense of the delight with which money was passing from the pockets of the buyers to that of the stallholders. The customers looked at the palaeontologist and me, wondering — I assumed — whether we might be a marketing ploy devised by the owner of the stall to lure in the public. Arsuaga continued with his speech, unaware of the curiosity we were arousing.

"Let's stay on track," I agreed.

"Imagine we brought a chimpanzee, a gorilla, and an Australopithecus male here."

I gave a discreet little smile.

"What's funny?"

"Nothing."

"No, what's funny?"

"It reminds me of those jokes about an Englishman, a Frenchman, and a Spaniard. I was wondering which one was the Spaniard."

"OK, very funny. Three primates. You get the idea? Three primates, one of them a hominid. Three links in the chain of evolution."

"I get it, pure Palaeolithic."

"Palaeolithic. Gorillas are folivores, or foliphagous. That means they eat leaves, greenery. They like the tender parts of vegetables. The gorilla, in the jungle, lives submerged in a sea of food. It consumes its environment, everything in the landscape around it, and for this reason never travels far, because the landscape is its food. But that which is abundant is also rather poor fare — it's low in calories, and you have to spend the whole day eating. What would the gorilla say to the people working on this stall?"

"Oh, I don't know, to give him all the green things."

"That's it: the lettuce, the spinach, the leeks, the chard, these endives, the curly ones, too … Everything that's green. Give me everything that's green, he'd say."

"Alright."

"Then along comes the chimpanzee, who's a frugivore. He would ask to be given all the ripe fruits, not the green ones. Of course, the chimp eats from the gorilla's store, and vice versa — the borders between them aren't hard and fast. But, fundamentally, the gorilla is a folivore, and the chimp, a frugivore. But there's no protein in fruit, only sugars and water."

"Sugars and water," I repeated, while raising my eyebrows in a what-can-you-do expression to the greengrocer, who had just thrown me his third questioning glance.

"Well," concluded the palaeontologist, "for now we'll go with that: the chimpanzee would leave the stall with several kilos of fruit, and the gorilla, several kilos of vegetables."

"Yes."

"And then along comes the Australopithecus. A primate, like the two who came before, but also a hominid. A notable evolutionary leap has already taken place. We're talking about a biped approximately one-and-a-half metres tall. Remember Lucy and the Beatles song."

"I remember."

"The Australopithecus would fill his basket with both things, fruit and veg, but this hominid's molars are larger than those of chimpanzees and gorillas, and these molars have a thick coating of enamel on them. That means, apart from fruit and veg, which come from the jungle, it chews other things, things that don't need to be cut up, meaning its rear teeth are more developed than its front ones. The chimpanzee, by contrast, has more developed front teeth, because a melon needs cutting up. The Australopithecus, who comes and goes in the jungle, has incorporated into his diet products that come in smaller, but higher-calorie units — hence the front teeth growing smaller and the rear ones bigger, with a thickening of the enamel. Are you following all this?"

"Something's changed," I said.

"Something's changed," said the palaeontologist. "What does he eat? He eats grains, legumes. Lentils and beans, for example. Fruits with skin, though the skin needs peeling. The jaws of the Paranthropus — the Australopithecus is a kind of Paranthropus — were regular nut-cracking machines. We're talking about biomechanics now, about the body as a machine.

In short, he would fundamentally eat the things you see conserved in jars: lentils, chickpeas, peas, beans ..."

"I get it," I said, setting off so we could go and draw attention to ourselves someplace else.

Arsuaga stopped me.

"Wait, we mustn't wander too far — there's a full explanation of our origins right here on this stall. You see, organisms have two missions: the economic one, concerned with having enough to eat, and the reproductive one."

"The whole be-fruitful-and-multiply thing from the Bible."

"Two missions. What do we need to carry them out?"

"Well ..."

"I'll tell you: proteins, because proteins are the building blocks of the body, but lipids, or fats, as well — which produce calories — and carbohydrates, which are the energy molecules. The body converts carbohydrates into glucose, and that's what the brain feeds on: glucose in its pure state."

A phone rang somewhere near us, and a lady answered it after some desperate rummaging in the depths of her bag. I sensed she was talking to her husband, and heard her telling him there were no mushrooms, even though she was standing right in front of them. "Well," she concluded, "next time you can come do the shopping yourself." Arsuaga hadn't even registered the conversation, because he was still doing his own thing, while I, while still listening to him, was occasionally throwing apologetic glances at the greengrocers and the customers whose way we were obstructing.

"For the chimpanzees," the palaeontologist said, "there's the incentive to start hunting, because, as I was saying before, vegetables are quite low in calories. Some groups of males go hunting for small monkeys, the babies in particular."

"They cooperate when they hunt?" I asked, amazed.

"That's a point of considerable debate. Do you believe that wolves *cooperate*, or that they all simply follow the same quarry? To my mind, there's no cooperation there; for cooperation in the hunt, everyone has to get a share afterwards. Anyway, this is one of the big questions in social biology, but I'm a sceptic. Cooperation requires a considerable degree of complexity."

"You were saying that chimpanzees hunt."

"They hunt small monkeys and infant herbivores. Monkeys yay big," — he gestured — "about a kilo or so. Which, in the economy of the body, doesn't solve anything, but it's like a sweet for a little child: a shot in the arm. Chimpanzees have a real soft spot for meat; they do love brains in particular. It makes no difference in the overall calorie count, but with such a delicious morsel at your disposal, you can grease a few palms, make political manoeuvres, forge alliances, obtain sex."

I noticed that the palaeontologist's tone had turned nostalgic, which was impacting my own state of mind. I thought about how we'd come from there, from those hunters of small monkeys that constituted the little sweets of their day. I could almost see myself and my family tugging on the arms of the unfortunate monkey, while it was still alive, yanking them out by their roots and bringing them to our mouths. Arsuaga's speech was rather hypnotic; it had the effect of transporting you to the periods he was talking about. When I came to my senses again, I still had a slight taste in my mouth of the monkey I'd just eaten in my chimpanzee state.

"He who has a monkey," the palaeontologist then said, "has something the others want. But let's go back to Australopithecus. Some of what they consume is beyond the confines of tropical forest, because Australopithecus, remember, goes out into the savannah — which is not grasslands. A lot of people confuse savannahs and grasslands, but we're not going to do that. There are grains in the savannah, because it's drier there than in the jungle, and there are trees and shrubs whose fruit is within the reach of this bipedal primate we call Australopithecus."

"Is there a clear border between the jungle and the savannah?" I asked.

"No, it's graduated. But in the savannah, as I say, you find grains, the manipulation of which requires dextrous hands. Chimpanzees and gorillas can't manipulate a pistachio nut, they can't pick it up, they don't have pincers. All primates, us included, have opposable thumbs, but the monkey's is ridiculous, on top of which it's far removed from the tip of the index finger, because they have such long hands. A chimpanzee's hand is a hook."

"To hang from the branches."

"Certain authors," Arsuaga continued, "take the view that this new kind of diet is adapted to human characteristics. Blackberries, mulberries — berries in general — are found on shrubs. We have gone, thus, from the world of the small fruit to that of the grain. You don't need big front teeth, but you do need pincers to manipulate it and molars to masticate it. And you don't need to climb a tree now, because the berries are at the same height as you."

"And what else does Australopithecus find when he leaves the jungle and ventures out onto the savannah?"

"Light! In the tropical jungle, not a single photon penetrates as far as the floor. They all get snatched along the way, because light is highly prized by the leaves of the big trees. The jungle is dark; it's a prison. Out in the grasslands, on the other hand, there's light. Australopithecus wants to see light."

While I thought about the discovery of light, Arsuaga turned toward the fruit and veg stall we'd been standing next to for an hour, and exclaimed enthusiastically: "Oh, how beautiful, a greengrocer's!"

"Yes," I said.

"Look," he said, "we were going to jump ahead to the Neolithic, but let's move on from the Australopithecus to *Homo erectus* instead."

"Just as well," I pointed out, "to bring a bit of order to things."

The palaeontologist turned toward me and said, rather displeased: "Hey, what do you mean, 'order'? This isn't some tale. If it's a tale you want, go read Genesis. Evolution isn't structured like a tale. There's no exposition, crux, and dénouement. Evolution is the world of chaos."

"In evolution, don't some things happen after others?" I asked naïvely.

"Sometimes not. I'm proceeding chronologically as a mere device, OK?"

"OK, OK."

"Right, so over 90 per cent of the calories ingested by humanity come from rice, wheat, potatoes, and corn. Four plants. An extra-terrestrial would categorise us as vegetarians. Fine, so does that mean Neanderthals were carnivores? Of course they were, because for three quarters of the year there were no vegetables. You get fruit at the end of the summer and in autumn, that much is true, and in prodigious quantities. Consider the acorn. Estrabón used to say we were an acorn-eating people."

"Do acorns keep?"

"Well, you have to turn them into flour and make pies. The number of acorns in autumn is tremendous. Pies would be eaten all year round; they were once the basis of the Iberian diet. But in prehistoric times, there were no mills. They hadn't worked out how to mill things or grind them up."

Finally, Arsuaga decided to move, and we walked off toward a disappointing poultry stall, because he was still on the subject of *Homo erectus* and he wanted to see something that was the result of hunting, of which there was scarcely anything.

"Time was," he said to the man working there, "that the partridges and pheasants would be hung up on hooks, with all their plumage. It was a nice sight."

"We're not allowed them now," the poulterer explained.

"And these here are wood pigeons?"

"I think so," replied the man, slightly ashamed at the lack of hunted produce.

"They'll be off to Africa now," Arsuaga said. "They pass over a couple of times a year — get them while you can. At any rate, *Homo erectus* didn't get many birds. They were difficult to hunt, compared, for example, with deer, red meat."

We remained at the poultry stand a little while, in free-fall toward disappointment, then Arsuaga took hold of my arm and pulled me toward him, saying quietly: "All of this is farm-produced. The last wild thing we have left is fish, and those only partly. Within twenty-five or thirty years there won't be any wild fish either — it'll all be farmed. When I was a student, it used to be said that the ocean was humanity's larder. Not anymore: it's running out; we consume too much. Think on this: 96 per cent of the total weight of mammals on earth is composed of us humans, cows, and pigs. See?"

"I see," I said, astonished.

"As for birds, the ones for consumption (chicken, basically) represent 60-something per cent of all birds. Where does that leave you? We human beings make up a third of the bio-mass of all mammals in the world."

Naturally, this left me feeling puzzled.

"Shellfish gathering," he continued, as we stopped outside a fish stall, "appears at the end of the Palaeolithic. It's a transition activity. In prehistoric times, animals came in big packages: horses, mammoths, bison … Not just big, but really big. You've got to get yourself a lot of mussels to come up with the same number of calories that you obtain from a horse. Smaller fare is a revolution. The leap from horse to limpet constitutes an economic, mental, and social revolution. Look," he said, glancing at his watch, because we had been going for a couple of hours by now and he was running out of time, "we'll finish up with geophytes."

"Geophytes?"

"It's a term a botanist came up with when he was describing the different vegetable biotypes. Geophytes are plants in which the perennial portion is subterranean. It includes bulbs, tubers, roots, and rhizomes. They sprout once a year. With geophytes, their nutritional content lies in the subterranean part. I'm referring to the starch — the carbohydrate, in short. This is how plants store energy: in the form of starch. A potato is pure starch. Wheat is starch; the same goes for rice. In geophytes, the starch is stored under the earth. You follow?"

"Of course, of course."

"Plants don't want to be eaten. The ones we're talking about bury themselves as a form of self-defence deep in the soil, far down where the little animals can't reach. What I want here is for you to end up with the concept of the geophyte."

"I'm working on it, believe me."

"They aren't such a big deal in Europe. But they are in

Africa, where we come from. Let's go back to the veg stall."

We returned, to the shock of the sellers, who thought they were rid of us.

"Here are the sweet potatoes, which by the way come from the Americas, the Andes. Many people have relied on this as a foodstuff, because it's so high in starch. Think about the potato, on which the whole of Ireland subsisted until the famine of 1845, which was brought about by a microorganism similar to a mushroom that caused a blight. It was the single crop there: they ate the skins, everything."

"When I was little," I said, "in Valencia, I had a lot of sweet potatoes. They used to roast them in the oven."

"And then, look, a cassava. Ever tried one?"

"I don't know, I don't think so."

"Now, listen: onions, garlic, leeks, asparagus, the potato, the sweet potato … Everything you see here is a geophyte. Cultivated geophytes, of course, but geophytes all the same. Why are they important?"

"Because they stop you being hungry."

"Because they are one hell of a resource, if you know how to get at them. And what do you need for that? You need a kind of inverse spear. A spear that points down instead of up. The inverse spear, or spade, is the symbol of Palaeolithic woman, just as the spear that points skyward is the man's."

"Right."

"We know how important geophytes are for modern-day hunter-gatherers. We have been able to establish that men spend their time hunting and the women gathering small items (spiders, insects, geophytes, et cetera). The men bring in half the calories by hunting, and the woman, children, and the older members of the group bring in the other half with geophytes.

With the inverse spear I was talking about, which is a spade, you can get down to the geophyte, if you become strong enough. That's what I mean when I say they are a resource that's very difficult for animals to get to."

"And Australopithecus, with their stature, could they get to them?"

"I don't believe so. But we can't be sure, because wooden spades would have rotted away. There's no fossil record of them. But by the time we get to *Homo erectus*, he was strong enough to get down to them. From *Homo erectus* onwards, you get this division of labour, of huge importance in the story of evolution because it expands the economic resources. Geophytes are a very regular and complementary resource. They mean the children have something to eat every day. Small items, as you see, provide a stable nutritional base. Small items have been, in terms of calories, as important as large game. The advantage of the small is in its regularity and predictability."

"Until what point were small things essential for human feeding?"

"Until the Neolithic came in."

"So the Neolithic revolution, it was a revolution by women?"

"Fundamentally, yes. Though you mustn't forget the elderly, who no longer take part in the hunt, and the children, who are yet to join it."

"So who invented agriculture?"

"Women, undoubtedly. The men are out there all day, tracking bison, horses, mammoths, the big units. Your typical macho man wants to come back home with the bison, because that signifies status, power. Prehistoric art shows the hunters coming back, and the children, the elderly, and the women awaiting their return. But I think the more usual scene would be the hunter coming back empty-handed and the women there with the geophytes, the limpets gathered down by the shore — the small items, in short."

"Whatever's predictable, whatever's constant."

"And that's where the management of resources comes in — it's just around the corner at this point — which places us one step away from agriculture. Managing resources already implies a high level of cognition. You have to understand the seasons, for example; to know where to be in spring, where in autumn. And be up to speed with the way the system functions, to make it work for you. You have to know which species to focus your energies on, and which not to. In conclusion, because I really do have to dash now, remember the distinction between favoured species and cultivated species. When you favour the growth or appearance of a certain vegetable, you aren't in the era of agriculture yet, but you're one step away."

"Wait, give me an example of a favoured species."

"Look, I'd say the acorn is in general quite bitter. It's possible that they'll find an oak tree one day that produces sweet ones, and they'll then tear out all the rest in order to grow those. It's a hypothesis."

I accompanied the palaeontologist to the exit, and outside the market there was a car waiting for him, at the door to which I said goodbye.

"We've still got the Neolithic left to cover," he said, winding the window down before the car pulled away.

Then I returned to the market to buy half a kilo of clams and three sweet potatoes.

"What's that?" asked my wife when she came into the kitchen.

"Small items," I said. "A late-Palaeolithic lunch. It'll be delicious — you'll see."

SIX

The marvellous biped

On 16 January, a light but icy breeze was coming in off the Guadarrama mountains — confirmation of the old saying about the wind in Madrid being capable of killing a man, but not of blowing out a candle. There used to be a time when people would line their torsos with newspaper to protect themselves from this gentle, deadly current.

It must have been just gone four in the afternoon, because I was taking my postprandial nap after my lunch in front of the TV when the phone rang.

"Juanjo," said the voice at the other end, "it's Arsuaga. I'm outside. Can you come down?"

I cleared my head a bit, went outside, and there the palaeontologist was, waiting for me, his expression somewhere between amused and mischievous.

"What's up?" I said.

"Are you busy right now?"

"Not right now, no."

"And are there any playgrounds nearby? There's something I want to show you."

I live in the Alameda de Osuna neighbourhood, right by

the Juan Carlos I Park, which is huge, even bigger than the Retiro, and which does have countless recreational facilities for kids within it. I went back inside to put on some warm clothes (inadequately, as will become apparent), and fifteen minutes later we were walking through the gates.

"What are we doing here?" I asked.

"We're going to watch children swinging around on ropes and playing on seesaws."

The palaeontologist does live in the real world occasionally, but only occasionally.

"There's no way kids'll be out in this cold," I said. "Besides, they're still in school at this time of day."

He looked surprised at this piece of information he clearly hadn't anticipated, but immediately reacted: "Let's stretch our legs anyway."

We started walking.

To our right, we could make out the snowy peaks of the mountains, and when I looked at his face I could see the icy breath being exhaled from his lungs.

"Cold, isn't it?" I exclaimed, hoping to encourage him to give up on whatever he intended.

"I wanted to talk to you about biomechanics," he said, ignoring my comment. "About the mechanics of the body, how bipedal locomotion works, and so on. Incidentally, I went to Amusco the other day, just near Palencia. It's the birthplace of the greatest Spanish doctor after Cajal, who's nonetheless completely unsung. Juan Valverde de Amusco. Have you heard of him?"

"Not really, no."

"A contemporary and follower of Vesalius. He wrote a book on anatomy that no doctor in Europe would ever leave the house without. I've got a facsimile. What about Vesalius, do you know anything about him?"

"Vesalius does ring a bell," I stammered.

"The father of modern anatomy. Died in the mid-1500s. Before him, we knew nothing about the human organism."

"We didn't know we had bodies?"

"The body, to this day, is a mystery to most people. A mystery. But I need a seesaw to explain something to you."

"There's one over there," I said, pointing at some children's facilities that were completely empty, as was only to be expected.

"It's tiny — let's see if we can find another one."

We went on walking through the silent, deserted park. The bare branches of the trees looked like arms raised to the heavens in terror. Some of them ended in leafless stems that looked like skeletal fingers, with their respective jointed bones. I was suddenly assailed by a feeling of being inside a Stephen King novel. This man had brought me here to kill me.

"Nowadays," Arsuaga continued, "the seesaws, swings, and slides are all very safe. The council makes it so the children can't smash their teeth in, like I did mine when I was little, because the parents would sue."

"People were less litigious then," I agreed.

"Do you not get any starlings in this park?"

"We've got blackbirds and ducks, I don't know — and millions of parakeets."

"Don't even talk to me about parakeets. I hate the things."

"But we were on Vesalius," I said, getting us back on track.

"Vesalius, yes. A phenomenon, a genius. He charted the human body, from head to toe. Until Vesalius came along, people were still relying on Galen's descriptions. But Galen never dissected human bodies, only those of pigs and monkeys.

It was Vesalius who first looked inside corpses."

"Leonardo da Vinci had forensic inclinations, too."

"But da Vinci had no feeling for anatomy. People always complain when I say this. I mean, he was a great artist, but no anatomist. As a scientist, he really was very poor. You look at the human figures he drew, and you say, perfect. That's until you take a closer look, and you notice it's missing this thing and that thing. Oh, and there's a few too many of this other thing."

"So until Vesalius, no one knew anything about the body."

"Not a thing."

"And other doctors in antiquity? Hippocrates?"

"They didn't have a clue. Hippocrates knew about medicinal plants, but he didn't have a clue about anatomy. Medicine and anatomy are different subjects. Anatomy's all about research, knowledge, while medicine's about how that gets applied. Doctors had practical know-how: how to perform incisions, how to sew people up, how to assist in childbirth."

"How to extract a tooth," I added, recalling an engraving from the period.

As we moved deeper into the park, the frozen breeze made its way through the successive layers of clothing I was wearing. Soon, I thought, it would reach my skin, then my muscles, and after them the bones, and later the marrow of the bones. When the cold got in there, into the marrow, there was no way you could shake it off. The silence, meanwhile, was such that we could hear our footsteps on the dry earth, until a cloud of three or four hundred parakeets flew shrieking overhead before coming to rest menacingly on the ground, very close to where we were walking. This only intensified my feeling of being inside a horror novel.

"Little bastards!" Arsuaga glared at the birds. "They'll be

the end of all the native species, because they're so difficult to exterminate. Write this down."

"I'm recording it," I said, showing him the dictaphone.

"But write it down as well, in that little red notebook you've always got with you."

I took out the red notebook and a biro.

"Go on."

"Sociability," he said. "The most successful species in the story of evolution are the sociable ones. That's what our success is down to. These bastards have communal nests, and they're very difficult to eradicate because they work together. As individuals, they aren't that robust; but as a species, just you try and take them out. The colonies, they're invincible."

"We were on Vesalius," I pointed out, trying to get the conversation back on the rails.

"Are you in a hurry?"

"Just so we don't wander too far off topic."

"You have got an incomprehensible fear of digressions. Relax! But yes, alright, we'll go back to Vesalius if it'll calm you down. Make a note of this, too: it's anatomy we're concerned with at the moment."

"And what does that mean?"

"That Vesalius is not about physiology."

"What's the difference?"

"Anatomy is structure. Physiology, the way it all works."

"And so, anatomy," I tried to clarify for myself, "would be the description of the organ, and physiology the function that organ plays within the body."

"Exactly."

"Kind of reminds me of the difference between morphology and syntax in grammar. Morphological analysis studies a word, and syntax the role the word plays in a sentence."

"Right, because all terms in linguistics come from anatomy. Morphosyntax, morphemes … Originally used in biology. Which is all to say, you can study the heart on a steel tray, or you can study the circulation of the blood. The heart, considered in isolation, is anatomy. Once you start linking it to circulation, you're in the arena of physiology. OK?"

"OK."

"But when it comes to circulation, you need a living person if you want to look at that, because blood doesn't circulate in corpses."

"Same with locomotion," I said, suggestively, hoping he might walk a bit faster, to shake off the cold.

"With corpses, all you get is anatomy, structure. Vesalius's followers were very innovative, compared with the Galenists, who were the old school, and the ones with all the power. They clashed regularly in the universities. There's an anecdote about some argument between a follower of Vesalius and a follower of Galen, with the former saying that Galen was wrong about something in particular — I can't remember what. How dare you go against the established authorities?, said the Galenist. Because I've checked it out on a human corpse, said the Vesalian. Well, said the Galenist, death's got it wrong."

"Which is more or less what the priests were saying about heliocentricity back in the day," I pointed out. "That reality was wrong."

"More or less. The thing is, Vesalius's followers started looking at what we're made of, and how. Muscles, bones, internal organs … This is going to help us in talking about the structure. The things that differentiate us from quadrupeds are from the waist down; our differences from the majority of land-dwelling mammals are from the waist up. From the waist up, we're more or less chimpanzees. From the waist down, humans. It's a strange combination, if you stop to think about it."

"Like a centaur."

"We're chimeras."

The palaeontologist does just come out with disturbing things like this occasionally. The chimera, in Greek mythology, is a fantastical monster, depicted as having the head of a lion, the body of a goat, and a dragon's tail. But it's also a term we use to refer to an unattainable desire.

"We are chimeras," I repeated out loud. And I added to myself, in tribute to Shakespeare: *We are such stuff as dreams are made on.*

"We're bipedal simians, not monkeys," Arsuaga said, as though correcting me. "The difference is subtle, but interesting. We are primates that belong to the simian category — monkeys without tails, that is. From the waist up, we're like all the other simians. Flat-chested, with rib cages that don't stick out in front. We've been squashed, do you see?"

The palaeontologist stopped and put one of his hands on my chest and the other on my back to help me understand the compression to which my organs are subject.

"It's barely a handspan between front and back, and in that handspan you've got the heart, lungs, et cetera. Quadrupeds, on the other hand, are *laterally* squashed."

"All of them?"

"Without exception. Think about dogs, and the shapes of their rib cages, and the way their shoulder-blades attach. Then compare them with the way yours are positioned."

I thought about a dog's ribcage, about the position of its shoulder-blades and mine. Arsuaga was awaiting my answer with a questioning look.

"Do you see, or don't you?" he asked impatiently.

"I see."

"OK, well put quadrupeds aside for now. We're going to consider the biomechanical efficiency of your body and mine. So, where is our centre of mass?"

"Are you talking about the centre of gravity?"

"Centre of gravity, centre of mass, whatever. The point where bodyweight is concentrated. Your centre of gravity is between your bellybutton and your pubis, in line with your belt buckle, but inside your body. If you go in a straight line down from that point to the ground, the straight line will end between your feet. Yes or no?"

"Yes."

"We'll call the area occupied by the two feet the 'support base'."

Now the palaeontologist stood behind me, and invited me to fall backwards while keeping my body straight, like in the trust exercise that people do in therapy groups.

"It won't hurt," he said. "I'll catch you."

I allowed myself to fall, thinking this was definitely the spot where he was going to kill me. But instead he asked: "Now what's happened to that line that went from your centre of gravity to the ground?"

"It's moved," I said.

"And where does it fall now?"

"Outside my support base."

"Meaning, if I let you go, you'd fall. Objects fall or tip over if you push them so far that the imaginary plumb line between the centre of gravity and the ground falls outside their support base — your feet, in your case."

"Which is why," I pointed out, remembering something I'd

read once, "the Leaning Tower of Pisa doesn't fall, because the line of gravity is still within its base."

"Right. In solid objects, the centre of gravity lies somewhere in the mass of them, but what about this: where is the centre of gravity when it comes to a hollow object? Like a ball, for instance."

"Where?"

"Not in their body, but in the centre of the sphere."

"That's so odd!" I exclaimed.

"Isn't it?"

I tried to imagine the immaterial point where an empty cupboard's centre of gravity would be located, and I deduced that it would coincide with that of its soul.

"But let's put this mystery aside for a moment," said Arsuaga. "I'm not saying it's the Holy Trinity, but it isn't far off. The important concepts to be clear about are the centre of gravity, the line of gravity, and the support base, if you're to understand the utter marvel that is bipedalism, in which the plumb line we've been talking about always remains within the bounds necessary to avoid falling over."

"Right."

"We were saying your centre of gravity is at the same height as your belt buckle. Now, the less your centre of gravity moves around when you walk, the more efficient the body will be from a mechanical point of view. Take a few steps, and keep an eye on how your belt buckle moves."

I walked looking down at my navel, and noticed that the buckle traced an almost perfectly straight line, parallel to the ground. My centre of gravity barely moved up or down, but nor did it move from side to side.

"Quite something, isn't it?" Arsuaga looked triumphant.

"Human locomotion is a wonder of biomechanical engineering. The net result is that it takes very little energy for us to move around. We are a species made for covering long distances; we are a walking species."

"Is that how we've come so far?"

"Perhaps."

"And we're unique in that?"

"Among the primates, yes. But, if you don't mind, I say we forget about the top half of the body, the trunk, and go back to the lower half."

"Let's," I said, joining him again so that we could head further into the soundless park (the parakeets had disappeared, perhaps because the sun was beginning to go down).

"Monkeys," said the palaeontologist, "walk around on the branches of trees. They are quadrupeds who walk around on branches, like rats. I don't know if you've ever had the unpleasant experience of seeing a rat walking along power cables."

"Yeah, when I was a kid."

"Rats move along branches as easily as they do on the ground. Certain quadrupeds are designed expressly for that. But because branches aren't horizontal, these quadrupeds we're now talking about — monkeys — have prehensile hands and feet. Their hands and feet have adapted. It's the hands that are different. A monkey walks along in the same way a dog does, but nobody's ever seen a dog walking along a branch. Get it?"

"Got it."

"OK, so there's one group of primates that are heavy and large. They don't move around by walking on the tops of branches, but by hanging from them."

"Right."

"Think about it: a primate suddenly stops walking along the tops of branches, and starts hanging from them instead. What does it need?"

I had no answer.

"It needs long arms and long hands, which act as hooks. They hang from these hooks, and as a consequence they're flat-chested. We need a children's playground with bars so you can see how suspended locomotion works."

"Fifty metres further on," I said, "there's one that's a bit tucked away. My children loved it when they were little, because it felt like a secret den."

We walked at a good pace toward the secret playground, which meant, to my own private displeasure, leaving the exit even further behind. With the setting of the sun, the temperature had dropped two or three degrees. Arsuaga gave no indication of feeling the cold, though his nostrils were red. Nor did he appear to be in any hurry. In fact, he stopped when a solitary bird flew over our heads in order to explain to me that it was a green woodpecker. I'd never heard of green woodpeckers, which up until that point I'd thought were sparrows. I then gently urged him on, and we soon found ourselves standing before some children's facilities half hidden by the undergrowth. There was a castle and two slides, and a long pole from which the children loved to hang with a bit of help from their parents. But at that time there were no children and no parents, there was only cold, as well as a gathering darkness, and silence — an absolute torrent of silence, because even the birds seemed to have fled the harsh evening. It was a frightening place to be.

"Suspended locomotion," said the palaeontologist, going over to the pole, "has a technical name. Write this down:

brachiation. I'll skip the etymology, because it's obvious, isn't it? From 'brachium,' Latin for 'arm'."

"Yes."

"Can you imagine a cat hanging from this pole?"

"Well, no."

"What about a dog?"

"No, nor that."

"Well, of course not, their hands aren't shaped like hooks. So, the only thing you can picture doing that is a gorilla or a chimpanzee."

"If a dog could hang," I asked naïvely, "would it become flat-chested, in time?"

"That is what's known as Lamarckism, a heretical perspective of which I hope to cure you. But not just now."

"OK."

Then he went and hung from the pole, before continuing: "This capacity explains our origins as tree dwellers, and the fact we belong to the evolutionary group of simians, which is the family we come from."

"If you get tired, you can get down from there."

"Why would I get tired?! OK, so apart from hanging, we have to go places. Let's imagine I'm a primate. Tell me, how does a chimp get from A to B when it's hanging from a branch?"

"I don't know."

"People *think* they move around like this, sideways, but that isn't right. The movement they perform is actually mind-boggling."

"Brachiation."

"And this is what it entails."

The palaeontologist let go with one of his hands and spun a hundred and eighty degrees, which placed him a metre or a metre and a half from where he'd previously been. After carrying out the same exercise in alternating directions, he dropped to the ground, panting slightly at the effort.

"It's a very difficult movement to perform," he said, "because it involves the arm, the forearm, wrist … You need to have suitable morphology, as well as enormous strength. A chimp could hang from this bar with one hand, and smoke a cigarette with the other, no problem. It doesn't take the same effort as it does for me. A chimp could hang there for hours chatting with you, cool as a cucumber."

"And the thumb plays no part in those movements at all?"

"None at all. To the point that they barely have four fingers. Only hooks. A chimp hanging from the branch is like you right now. Am I right in saying the fact you're standing up isn't really in your thoughts?"

"Well, it is slightly, because I'm kind of tired, and also it's really cold."

"Don't be silly," he said, with some vehemence. "You don't even realise you're standing up. You aren't aware of it. Shame there aren't any children around, because children are still brachiators. If they were here, you'd see it more clearly."

"But it would be a bit strange having two older gentlemen standing here watching the children. We'd probably end up spending the night at the police station."

"Good, very good," said the palaeontologist. "We got as far as being simians. Yes or no?"

"Yes."

"Well, now we're going to be bipedal simians. Now we have to stand up. We've got ourselves one part of the body, the upper half; we've managed to make that part human. We've got an evolutionary understanding of the upper portion of the body.

Now it's time to stand up."

"Alright."

"As I was saying, we are designed for walking; we are a walking species. The longer the stride, the more efficient. We absolutely must find a seesaw."

"Let's go to the one we saw when we came in, but let's hurry — it's already getting dark," I urged him, eager to escape that sinister part of the park where I was sure all the neighbourhood's dead children gathered after nightfall.

"Up until now," he said as we moved off, "we've been talking about anatomy. Let's go back to biomechanics. Do you remember studying levers when you were an undergrad?"

"More or less. Give me a firm place to stand and I can move the world — that's what Archimedes said."

"Well, you get three types of lever: first-, second-, and third-grade ones."

We were walking round the edge of the park's lake, over whose watery pane a very fine blanket of fog was beginning to settle like a shroud. I spotted a duck in flight, and I said: "Oh, look — a duck."

"No, dammit, that's a cormorant," he said. "A cormorant's not something to mess around with. This park's fantastic."

"They say it's a new concept for a park: an intelligent park, as if all the old ones were really dumb."

"I could tell you a few things about the cormorant, which is wonderful at fishing. In some countries, the fishermen lodge something inside the cormorants' throats so they can't swallow the fish. Then they go and steal the fish they've caught."

"OK, but let's not get distracted by the cormorant."

"Again, that fear of digressions. Where's that trauma come from?"

"I don't know — I like to get straight to the point."

"Levers," he said, resignedly. "We were talking about levers. Mechanics is all about different kinds of lever. When it comes to machines, everything that isn't an engine is a lever. A seesaw is a first-grade lever."

"And what have a seesaw and a lever got to do with the body?"

"We were saying before how useful it is for the body's centre of gravity to move forward in a straight line, parallel to the ground. Keep in mind that we're able to expend two-and-a-half thousand calories a day, more or less. It's no easy thing getting our hands on those calories, and it's worth our while to manage them well. The solution being that the centre of gravity barely moves about while we walk. There are two kinds of movement that would be no good at all: an up-and-down movement, and a side-to-side movement. Because we are bipeds, when I lift my right foot, for example, it produces two forces: first, gravity, which pulls the body over to the unsupported side. But so that I don't fall over, I lean slightly to the left, above the leg that is supported on the ground, with that hip producing a counteractive force. Meaning that, if you think about it, walking is tantamount to constantly falling over."

"Amazing!", I exclaimed, genuinely marvelling. "Walking is being in a constant state of falling, just the same way that to live is to be perpetually dying."

"But you fall in a controlled way," the palaeontologist added. My rhetorical discovery hadn't even registered. "Such that you don't realise. You don't notice you're falling."

"You don't notice you're dying either," I said, though again it didn't seem to register at all.

The palaeontologist stopped to give me a practical demonstration: "See? I raise my right leg, but I don't collapse, I don't fall over, because of some muscles in this other leg, at hip height, which provide a counterbalance."

"Which muscles?"

"The abductors, which, when I raise my right leg, pull to the left, though not so much as to move the centre of gravity excessively. They apply a subtle correction, let's say, expending minimal calories. Again: we only get to use two-and-a-half thousand calories in the entire day, and that includes everything the brain needs, and the organs, and for the regulation of body temperature."

"All that on only two-and-a-half thousand Euro?"

"Not Euro. Calories."

"Sorry, it just popped out. I was thinking in accounting terms. So what you're saying is, we have an incredibly rigorous administrator inside us."

"It's called natural selection."

"And what did all that have to do with levers?"

"A lot, because the comportment of the body, in the shape of the movements I've just explained, is that of a first-grade lever, and as it happens, that's what a children's seesaw is, too. Which is why we need a seesaw."

"We're nearly there."

"Now, what qualifies as a first-grade lever?" he asked.

"You tell me," I replied.

"They have the fulcrum in the middle."

"A fulcrum?"

"The needle. The needle over which the scales tilt, I mean. With a seesaw, the needle is placed in the middle of the two forces. If you sit at one end of the seesaw and I sit at the other, each of us represents a force. If I weigh more than you, the seesaw dips down on my side. What can you do to prevent that?"

"Ask you to please not take advantage," I said.

"In terms of *mechanics*, what option do you have?"

"I … can't think of any."

"If I've raised your end up, there's nothing you can do, except to ask the engineer to move the fulcrum, or needle, whatever you want to call it, closer to my end. That means you have more lever, more arm, and as a consequence more force. Yes or no?"

"And if we translate that to the body?"

"What evolution does is to play around with arms of the lever. The fulcrum, the pivot point, is the articulation of the hip with the ball at the top of the femur. Remember, the centre of gravity is at the height of the belt buckle. When you lift one leg to take a forward step, how does the body avoid tipping over to the side that's sailing through the air? Bear in mind that the centre of gravity is not directly above the foot that's on the ground, it's at the plumb line, which is now off to one side. If you lift the right leg, shouldn't you then fall over to the right side, given that you're subject to gravity? That's where the muscles called the abductors come into play, pulling in the opposite direction to keep you level, offsetting gravity. It's like on the seesaw. You, who surely weigh more than I do, are the force of gravity that wants to make the seesaw tip over to your side, and I am the abductors trying to keep the seesaw horizontal. OK, so the longer your lever is, the greater your advantage. You want to be as far away as possible from the seesaw's pivot point. As do I, for the same reason.

"In anatomy, the abductor muscles' lever arm is the femoral neck, and it can't be too long, otherwise it would be in danger of breaking. People with osteoporosis often break their femoral necks. Which means I can't go all that far away from you on the seesaw, not as far as I would like. That being the case, what I'd really like is for you to come nearer — but that could only happen if the birth canal became narrower, which would make childbirth impossible. The evolutionary upshot is a compromise, one result of which is that women have a hard time in labour. And, speaking of gravity, you shouldn't confuse mass and weight. My mass is all the material of which I'm composed, and my weight is the force exercised by gravity on that mass. On the moon, my mass is the same as it is here, but my weight changes."

I lowered my leg and looked around, just in case there was anybody watching these two adults, one of them standing on one leg and the other with a hand supporting his hip. But there wasn't a soul about, and the sun was poised to disappear completely on the western side of the park, behind us.

"All these movements are so synchronised that the displacement of the centre of gravity is utterly minimal. People go to doctors with back problems, and some are told, 'Of course, if only we hadn't evolved to become bipeds, if only we'd carried on being quadrupeds ...' This, in evolutionary terms, is nonsense, because we'd be quite useless as a species. Or it's like saying every other species is fabulous except ours, which is rubbish. So ingrained is this in the heads of some evolutionary biologists that we've spent years looking for some advantage that could compensate for this supposed biomechanical disaster. We'd become so used to the idea of ourselves as, biomechanically speaking, a botched job that everybody was always saying there must be some advantage to make up for a botched job like this."

"And did they find it?"

"Well, they found lots of things: like the idea that bipedalism was good, because it meant you could have sex facing your partner, or for being able to juice oranges, for being able to give commands at a distance — goodness knows, everyone had their own answer. But there's no need for any explanation. Bipedalism is just a marvel."

At this point we arrived at one of the children's play areas near the exit, where the small seesaw was located that the palaeontologist had spurned at the start of the evening, and which he was now determined that we should get onto. It was already dark, and our shadows, lengthened by the light from the streetlamps, spread across the sand where the kids commonly rolled around.

We were a scary sight.

Here we had two older gentlemen, who looked like they'd escaped from a madhouse, seesawing amid this frozen Wednesday in January. As we rose and fell, the palaeontologist demonstrated more than sufficiently that it only took one of us to move a little way forward or backwards in his seat for the forces to become unbalanced. Next, continuing to propel himself upwards and to fall back down, he launched into an extremely extensive tribute to the femur, which drifted away into the air like teardrops in the rain, because I couldn't make notes or catch it on the dictaphone. I was having to deal with too many things at once. I heard him say that that particular bone was one of evolution's great inventions.

"A real architectural feat," he added. "Even the best engineer in the world couldn't have come up with something like the femoral neck, which supports our entire bodyweight as we walk along."

I remained alert to the shadows, just in case somebody I knew from the neighbourhood happened to be taking their dog for a late walk and surprised me in this scene, which looked like something out of a gothic novel. When I finally managed to get him off the seesaw and guide him to the exit, he took a pocket watch out of somewhere or other and put it down on the ground.

"Stop for a moment and look at this watch," he said.

I stopped and looked at it.

"If it wasn't a watch but a stone, and I asked you what that stone was doing here, you would tell me it's always been here, wouldn't you? That it's part of nature."

"Of course."

"But if it's a watch rather than a stone, you'd say someone must have put it there. Yes or no?"

"Yes."

"Well, Darwin's entire career was an attempt to show that the watch could have made itself. The idea of the watch on the ground comes from William Paley, a utilitarian from the 1700s."

"The one who called God the watchmaker of the universe?"

"The very same," said Arsuaga, picking it up. "All of Darwin's work is an attempt to challenge that idea of Paley's. We'll talk more. I need to dash now — I'm late for a Pedro Guerra concert."

The following morning, I quit my bed at around six o'clock, as is my custom, but I could tell at once that something was wrong, like when you accidentally get into your neighbour's car, which is identical to yours and which mysteriously also opens with your key fob. You're already sitting there, just about to start the engine, but you pause because something isn't fitting quite right. Maybe the seat isn't the usual distance from the steering wheel. Maybe the inside of the car smells like somebody else, or the dashboard is cleaner than yours …

There's a syndrome, the Capgras delusion, which entails you getting out of bed one morning and going to the kitchen, where your parents are having breakfast. Except that they aren't your parents. They've been replaced during the night, though you have no way of proving this because these people are identical to the ones who kissed you goodnight when you went to bed. Nor are your siblings, who soon join proceedings, actually your siblings. It's possible that this very kitchen, being the same one as ever, does strike you as an exact copy of the regular one. A bad business, because there's nothing for you to do but pretend you haven't noticed the swap. The alternative would get you sent straight to the madhouse.

The person who quit his bed at six o'clock that morning was a guy identical to me, but it wasn't me. I went into the bathroom and had a quick wash, certain that it was the other guy who was washing himself, the person who'd supplanted me in the night. This other man was excessively aware of his own body. He was aware of how many biological and mechanical resources were set in motion each time he took a step toward one part of the house or another. This other man saw himself as a machine, as a robot of unusual perfection — had it not been for the distress signals that its joints were sending to its brain.

It didn't take me long to work out what was going on: I had a fever.

Me, *I* had a fever — not the other guy. I checked on the thermometer: 38.5. Too high for that time of day. The other symptoms gradually appeared, until the completion of a picture that the doctor, when he showed up with his stethoscope, classified as pneumonia.

Curled up in the bedsheets, subjected to very aggressive medication, my mind was occupied with images from the previous day at the Juan Carlos I park. I saw the cormorant, the parakeets, the green woodpecker, but most of all I saw myself trying to understand the basic mechanics of bipedal locomotion, and I felt a certain apprehension at the thought that, when it came down to it, this was all we were: an artefact perfectly designed to cover great distances, to screw face-to-face, and to squeeze oranges. I recalled Descartes and his ideas around the animal-machine I'd learned about in my years studying arts and philosophy. Descartes' dog. The animal as mere artefact.

At that moment, as if Arsuaga, from wherever he might happen to be, were reading my mind, the phone on the bedside table rang, and it was him, of course. I didn't admit I'd come down with something, so as not to seem like a weakling.

"What's up?" I said.

"I've been thinking about it, and I believe my explanations yesterday might have tended a bit toward the mechanical."

"A bit, yeah."

"I possibly failed to give the proper context. I talked to you about the first-ever anatomists and physiologists. The world of the sixteenth century, which is when the Baroque scientific revolution came about, was all about machines. Everything was explained in terms of mechanics. Galileo was the first, but then came Descartes, Newton, Leibniz, et cetera; they all started to realise that everything could be boiled down to mechanics. The world was a machine; the human being, too. The body was an automaton, hence their fascination with putting automata in cathedrals, like the *papamoscos* clock in Burgos."

"Is fever an automation?" I asked.

"What's fever got to do with it? What I'm trying to say is that the 1600s is the century of physics, in the way that the 1700s is that of chemistry; the 1800s is the century of biology; and the twentieth century is all about psychology. We spent a lot of yesterday in the 1600s, the century of physics and hence of mechanics. It's a century in thrall to mechanics."

"Right," I said.

"Is something wrong?"

"No, nothing."

"Of course, Descartes also believed in a dualism of matter and spirit. In fact, he located the soul in the pineal gland, and actually that wasn't a million miles away, given that it's an endocrine organ."

"And what use does a mechanicist have for a soul?"

"Somewhere, he or she says, there must be something that thinks. There must be a coordination point between the machine and what Descartes called, in opposition to the body, the *res cogitans*, or thinking substance."

"He didn't simply accept our just being machines?"

"No, the body was missing something, and this duality was his way of solving it. The physicists didn't believe in the God of the Bible, the bearded guy you pray to when you want to pass an exam, but they suspected something was up. They spent their lives wondering if there was anybody there. Is there anybody there, is there anybody there?"

"And the biologists?"

"We biologists see things being born, growing, reproducing, dying, and finally rotting. Biologists who believe in God are few and far between, but the physicists and the mathematicians, they can't stop wondering what's going on. What's going on that means everything functions with the precision of a machine, with a language that can be represented with very simple equations."

"And what is going on?"

"That's what we're going to see."

At this point I had a coughing fit that made me almost spit up my lungs. I covered the receiver so the palaeontologist wouldn't hear it as he went on talking: "But I want you to pause on Paley's watch, because, as I said yesterday, all of Darwin's work is aimed at proving that the watch has created itself."

"How is a watch going to create itself?" I said, once I'd recovered.

"I'm going to take you to a place where you'll be able to see."

"What place?"

"It's a surprise."

"How was the Pedro Guerra concert?"

"I was late, because of you."

SEVEN

Refounding Bettonia

"The greatest thing you can aspire to, if you can't be a Basque, is to be a Celt," the palaeontologist said to me as he changed lanes with a rather sharp turn to the wheel of his Nissan.

It was 8.00 a.m. on 26 March, and we had once again skipped school. The occupants of the cars surrounding us on the A Coruña motorway were off to work: you needed only look at their expressions of dejection or rage. Sometimes, if you kept your eye on one of those faces for a little while, you found that they were smiling to themselves: they'd just imagined that their boss was dying or that they'd won the lottery — that life, in short, was going to start treating them the way they deserved.

The temperature outside the car was two degrees, but the sun was shining. It was forecast to reach fifteen by noon. There were generalised complaints about how long spring was taking to arrive.

"What do you mean by that, the thing about not being able to aspire to anything greater than being a Celt?" I asked after putting the heating up a bit, which I did despite a critical stare from Arsuaga, who never feels cold (or hot, I think).

"Look, it's great being a Celt. What's left to live for if you aren't a Celt?"

"I don't know, you tell me."

"Working in an office, going to the supermarket, picking the kids up from school ..."

"Some days I pick up my grandkids."

"And that's all well and good, but a person needs something else — needs to *be somebody*. Just picture it: us, Celts! The problem of existence, solved. We come together and form a great nation. The whole nine yards."

I reflected for a few moments, and finally acknowledged that he was right: "It's true, for some time now, everyone's wanted to be Celtic: the Galicians, the Asturians ..."

"And the Cantabrians. Celtism induces strong emotions."

"Celtism and bagpipes," I ventured.

"They also play bagpipes in Turkey," he said, admonishingly.

I didn't know where he was taking me, and nor did I go so far as to ask because I'd started to find his unpredictability enjoyable. The only problem was that I hadn't had much breakfast, and when I don't have much breakfast I start thinking about lunch several hours too early.

"The Celts," Arsuaga continued, "occupied the north-west quadrant of the peninsula."

"And what's left of them now?"

"You and I are what's left of them. We're going to refound a nation today."

"What nation is that?"

"Bettonia. People spell it different ways, but you should do it with a 'b' and two 't's, which I think is the best way."

"Sounds very central European with those two t's."

"Well, be sure to include them. The Bettones — write this

down, too — are a Pre-Roman people who lived in the Sistema Central mountain range. What would go on to become the provinces of Ávila, Salamanca, and Cáceres. That was where the Bettones lived, a most tremendous tribe, one that you and I are going to make a nation of."

(Whoever said palaeontologists don't have a sense of humour?)

"And what do you need to build a nation?"

"A defeat, and a national cuisine. Any nation that hasn't suffered defeat isn't worthy of the name."

"Who was it that defeated them?"

"The Romans."

"And what was their national dish?"

"*Patatas revolconas.*"

The reference to potatoes perked me up, though I didn't know what this particular dish involved.

"What century are we talking about?"

"The fifth B.C.E. Which is where we're headed now, in our twenty-first-century car."

"And how come they were eating potatoes in prehistory, when the potato only came over from America?"

"Look, I know the potato didn't come along until centuries later, but I just can't imagine the Bettones eating anything else. Don't mess up my story."

"That's very rigorous of you. And were they a happy people?"

"Absolutely. All of these legendary peoples were happy, and democratic. They had assemblies — everything was decided by a show of hands. They had plenty of animals to hunt, and fish, everything in abundance."

"Were they already farmers and ranchers?"

"More than anything, they were farmers. And ecologists. The greatest ever, in fact, hence the need for this crusade of ours, in order to rehabilitate them."

"Where'd they come from?"

"Well, from central Europe, certainly. They spoke Celtic languages, or other languages that, if not Celtic, were Indo-European. Please focus, Juanjo. Fifth century B.C.E., that's where we're going."

And we went fast, since the moment we were outside Madrid's ring road and its traffic jams, the Nissan raced like a cheetah toward Ávila.

"Did they come over in large numbers?" I asked.

"No, don't think about this in terms of big migration flows. Certain élite warriors would have come along, taken control of the territory, and an aristocracy would have formed from there. Great to be a Celt, huh?"

"Hmm, I don't know."

"As for the potato anachronism you've tried to ruin my day with, I'd like to remind you that in *The Legend of Juan de Alzate* — my favourite book by Pío Baroja — the Basques are eating corn before the arrival of corn from the Americas. Baroja justified this by saying that he simply couldn't imagine his forebears eating millet, like they were canaries. So, they ate corn, and bugger the rest."

"The same buggery that meant the Bettones were eating potatoes before potatoes were even discovered?"

"Well, exactly. Nobody's perfect. The landscape you're going to see when we get to Bettonia is the same as the one they would have seen every day of their lives, because it hasn't changed one bit."

"And what else should I write down, except that they were

a happy people, cheerful and confident, and that they wanted for nothing."

"That they were often starving; these things aren't mutually exclusive. And real bastards, too. They spent their time robbing their neighbours, though it was never personal."

"It was business."

"Just the way things were. They lived in a mountainous area, which had valleys but wasn't great for agriculture. Very stony; bones littering the place; nowhere to get a plough in."

"They'd have had pastures."

"Pastures, of course, for the livestock. Every now and then these Bettones, and their friends the Lusitanians, would go pillaging along the River Duero and the River Tajo, stealing the cereal grown by the good folk along there. Whatever they could get their hands on. It was a way of getting through the lean periods, and helped to smooth over social distinctions."

"What else did they have?"

"Pigs. Cows and pigs, more than anything. Livestock is something you can accumulate. With accumulable capital, social stratification appears. The haves and the have-nots. In a clan system like this, you get haves and have-nots. Meaning that the have-nots are sometimes obliged to go out —"

"… pillaging."

"To keep themselves fed. Like the Vikings did. Meaning, in turn, that all the surrounding tribes were pretty pissed off with them, and when the Romans showed up, they asked for their help."

"The Romans were the Magnificent Seven."

"Something like that. But with taxes."

"What religion did the Bettones practise?"

"That we don't know. But the principal god of the Celts was

Lugh. Lugh spawned a lot of places names."

"Like Lugo?" I ventured.

"For example. Lugh would come to be like Thor for the Germanic peoples."

"Oh, look — cows," I said, pointing at a group grazing to our right.

"Ávila cows. The Carpetani lived in this area. They got nothing."

"Carpetani and Bettones. Is that where the Carpetovetonic mountain range got its name from?"

"Exactly. You're getting your bearings, it seems."

"I'm not sure I like the landscape."

"The landscape of a place is the first thing we have to go on when trying to understand the past. Geography is all — it defines everything. The ports, the mountain passes. For starters, it dictates the distribution of the people. What we call the 'empty Spain' phenomenon is a product of geography. But there are two symbols that characterise the brave and noble Bettone."

"Potatoes," I ventured.

"And *verracos*. Potatoes and *verracos*, which are zoomorphic stone sculptures."

"The Bulls of Guisando?"

"For example. Where there are *verracos*, there were Bettones."

"And why do you like the Bettones?"

"I like pretty much everything, but today it's the Bettones' turn."

We had left the Nissan in the middle of nowhere to continue on foot up a steep slope. The temperature was a merciless three degrees. Behind us lay the Amblés valley, a magnificent depression through which the Adaja river flows. That meant we were right in the very heart of the province of Ávila. The bottom of the valley, which is wide and flat, conveys a feeling of harmony to the visitor, as if the fresh morning breeze (cold, more like) were carrying invisible particles of opium that numbed the pain of living.

"It's so nice, this whole skipping-school thing!" I exclaimed, overflowing with a euphoria that is quite unusual in a temperament such as my own.

The palaeontologist nodded his agreement, turning inward again, to a place where he had withdrawn without my noticing. When Arsuaga lapses into one of these absences, his face takes on an expression of nostalgia very like melancholy. He doesn't say anything, but I know he's mentally reconstructing a prehistoric scene. He is capable of seeing how a group of Bettones is making their way up from the very bottom of the valley. And indeed, at this very moment, he turned to me and said: "They would have driven their livestock right up here, into the fort."

He was referring to the Ulaca hill fort, which stands at the top of the mountain we had started to climb, and whose walls were perfectly visible from our current position.

"The fort," he noted, "is hearth and home, it's safety. Our familiar gods await us there, along with the other members of the clan. In the fort, you're safe from everything."

"Let's storm it, then!" I encouraged him, feeling in need of that mythical safety he was talking about.

"Before we do," Arsuaga said, "you need to feel the vertigo of losing it."

Then he explained to me that we were going to climb toward the right, moving away from the fort so as to experience the sensation of loss and, later, of rediscovery.

"As an aside, I'm going to show you something surprising."

We started the climb in total solitude. At one point, a fly

went past my face, dissolving into the atmosphere, which was as cold and abrasive as steel. Although transparent like glass.

"A fly!" I exclaimed.

"What is it?" asked Arsuaga, up ahead on the uneven path, without turning his head.

"Nothing — it was just a fly going past."

"Right," he said.

We could hear birds singing, but we couldn't see them — not a single bird, though they didn't stop yelling at one another. We passed a group of cows that the palaeontologist considered with a certain amount of contempt.

"These are the conquerors' cows," he said. "They're Charolais cows. Hence the reddish fur. Ours are the black Ávila ones."

"Right."

I saw that as we walked, the fort did indeed get further and further away, though part of its walls remained in sight. I started to feel some unease. There could have been wolves around there, or maybe wild dogs.

"Wouldn't it be best for us to go back?" I asked. "There's nothing of interest to see here."

"What do you mean, nothing of interest?" Arsuaga said, stopping. He looked angry. "Doesn't this granite landscape speak to you, all these rocks that have been waiting for centuries for us to come and visit them?"

I said nothing, not wanting to make matters worse. The palaeontologist was taking off his jumper, leaving him in just his T-shirt, despite the temperature still being low.

"Do you know where granite comes from?" he asked me.

"At this exact moment, no idea."

"It's igneous rock. Think about it: igneous. Formed by the cooling of volcanic lava."

The truth was, if you thought about it, it gave you the shivers. All those sculptural shapes, which by night must look like giants, were once a burning liquid.

"Have you noticed," Arsuaga continued, "that very often at the top of a vertical rock, you get a horizontal one?"

"That's true," I said.

"They're known as perched or balancing rocks, for obvious reasons."

We kept on climbing through the solidified granite landscape, never fully losing sight of the fort, though it became gradually harder to make it out. The feeling of solitude was such that, if you had told me we'd been dropped onto an alien planet, I'd have believed it. All of a sudden, as we turned a bend in the path, the wall disappeared.

"You can't see the fort anymore," I said, as though warning of a danger.

The palaeontologist stopped, slightly breathless, and invited me to walk on a few more metres, to the next bend, where he stopped again. He said: "We've entered the territory via a cattle drove that's been in existence for centuries. The fort, the settlement, safety, and security — we've left it all behind. Now, if we were to cross this sierra ahead of us, we'd come out at Talavera de la Reina."

He just looked at me, as if waiting for me to say something, but I had no interest in getting to Talavera de la Reina, if that was what he was suggesting, so I just raised my eyebrows questioningly.

"Do you not see anything remarkable around here?" he said. I scanned around us, while he continued: "Bear in mind that the most incredible phenomena tend to be very humble in appearance."

At this point, I saw on one side of the road, exactly on the curve we'd just taken, a vertical rock, about two-and-a-half metres high, wide and flat on top, on which a whole host of little stones had been deposited.

"That?" I said.

"That," he said. "That's Celtic. It's called the *Canto de los Responsos* — the Prayer Song for The Dead — because whenever somebody places a stone on top, a soul is released from purgatory."

"But the idea of purgatory comes later than prehistory."

"Of course, because it was Christianised later on. But in its Celtic origins, it was a rock of protection. You find these rocks on the routes and paths people would have taken. We're at a point here where the protective presence of the fort, of the hearth, is no longer sheltering us. We've entered what the Romans would have called the *saltus*. They distinguished between the *agro*, cultivated lands, and the *saltus*, which is unfamiliar, the forest, the wild, anything that hasn't been tamed by humans. The *saltus* is where the spirits live. There are dangers here, souls of the dead, and deities that don't belong to the domestic realm, and you'd do well to keep in their good books. These kinds of holy rocks, sacred rocks, have many names. Then, when everything was Christianised, they came to invoke souls in purgatory. But originally they represented the presence of spirits in the wild, ones you needed to propitiate. These rocks mark the border between what has been subjected to our control and what hasn't. They're somewhat reminiscent of the Greek *omphalos*, which is the place, also represented by a stone, where the underworld, the terrestrial plane, and the heavens meet. This is an omphalic site, no doubt about it."

I thought of the Borgesian Aleph, the point of confluence

for everything in the universe, but I said nothing because, after what Arsuaga had just said, the place where we were standing was thick with silence. Though caught up in this invisible timeless morass, we were nonetheless able to feel the trembling of the earth over which our Bettone ancestors would once have ridden. (They were good horsemen.)

We come from the place where we had just arrived.

When a few minutes had gone by, I said: "We should each toss up a stone to get a couple of souls out of purgatory."

The palaeontologist roused himself again.

"Don't even think about it! We have to leave it exactly as it is, out of respect."

When we had been retracing our steps for a little while and the fort appeared once again in view, Arsuaga, who I noticed had been grief-stricken ever since we had left the omphalic place, stopped.

"Perhaps we should have put those two stones up there," he said. "I'm feeling tortured by the idea that two souls are now stuck in purgatory because of us. Let's go back."

The idea of yet again losing sight of the fort, which had become my crucial reference point, terrified me, so we had a small argument in which I managed to persuade him that he had been assailed by a spell of a superstitious nature.

"You're right," he said finally. And then, looking at his watch: "Plus we need to make it up to Ulaca, and it's getting on."

The ascent to the Ulaca fort proved simultaneously stimulating and gruelling, as the gradient was very pronounced. I tripped a couple of times, and grazed my knees and the palms of my hands when they broke my fall. Arsuaga didn't see this happen, or pretended not to. The temperature had risen (four degrees, according to the app on my phone), and I was starting to sweat from the effort; but when I shed my jacket, the breeze passed through the sweater and chilled my sweat, so that I put it back on again and took it off again, by turns, until we reached the top of the mountain, with me a few metres behind the palaeontologist, who was waiting for me, panting slightly.

"Look at the valley from here," he said. "See how beautiful the Ávila sierra is. On the other side of it are the arable lands watered by the Duero."

I looked at the valley, and he was right — it was like something out of the imagination. It was as if we were inside a hyperrealist painting (hence its fantastical quality, too) produced by a seventeenth-century Dutch landscape artist. I wondered why we're condemned to experience the best moments in our lives as unreal.

It was eleven-twenty in the morning when we entered the fort by the same opening in the wall that was used as a gateway by our Bettone ancestors. Indeed, we had walked across the cobblestones that their sandals once trod (if they actually wore sandals — I must find out).

"The fort," said the palaeontologist, "is a mythical place at which one never actually arrives, but we've taken a shortcut through reality, and here we are. Who would've told you that yesterday?"

With our breathing back to its regular rhythm, we wandered unhurriedly through the inside of what had already become our home.

"Were there streets?" I asked.

"No, just scatterings of huts. Look at this green lichen here."

I crouched down beside a granite rock in the shape of a skull.

"It's Rhizocarpon geographicum. It got the name because it makes shapes that look like maps."

"It's true!" I exclaimed in surprise. Perhaps there's some dimension of reality where the countries shown in these maps actually exist."

"Perhaps."

"And what became of the valiant and noble Bettone people?"

"They were Romanised, diluted, liquidated, I don't know. Rome destroyed the syntax of clans in order to pave the way for cities. What do you think is better: being a part of a clan, or a city?"

"There are benefits to both," I said, noncommittally.

"Citizenship is great," he said, "because you start to become yourself. Within the clan, like in an anthill, the individual is the group. The Romans created the state. The state is responsible for building roads, it keeps you safe, gives you an identity. Ortega used to say that civilisation consists of taking a village and putting a hole in the centre: this hole is the public square, the forum. The forum turns its back on the natural world; it's a public space that is completely urban. In the forum, you see the beginning of thought, communication, politics, the market, the economy. It's the negation of nature; it's the non-countryside. The first thing to ask yourself about a culture is whether it has public spaces. If it does, then you're dealing with a civilisation in the modern meaning of the word. And if not, what you're looking at is merely a loose association of people."

At twelve we were still wandering aimlessly around the inside of the Ulaca fort walls. We were alone and a little out of it, as if we'd taken some hallucinogenic substance or other, or, failing that, a couple of ibuprofen. The wind at this altitude crashes and roars as if it wants us out. Suddenly, next to the wall, we happened upon some stone steps leading to an architectural feature in granite that was simultaneously simple and sophisticated. It was a prehistoric sacrificial spot.

"I'd say we're in about the second century B.C.E. now," said Arsuaga. Then, pointing to some grooves in the stone, "Blood from a sacrificed animal would have run off down here. Remember Lugh, the god."

"I remember."

"He's one of the old gods, out of prehistory, the kind who only asked human beings for reverences and sacrifice. Do you know when it is that the meddler god appears — the one who concerns himself with what we do, who punishes you if you sleep with the wrong person, and keeps a close eye on everything you think, given that it's possible to sin in thought as well as deed?"

"When?"

"I'll explain in the car, on the way back. For now, let's start heading down, then we can go and eat some *patatas revolconas* in homage to our ancestors."

On the way down, I slipped a couple of times in the same places I had fallen on the way up. But I kept moving, cheerful and confident, like a hungry Bettone, thinking about food.

"If we do refound Bettonia," I asked the palaeontologist, "what will we live on?"

"Subsidies, surely?"

We had lunch in Solosancho, a village in the valley, where, naturally, we ordered *patatas revolconas*. I didn't say anything so as not to cloud the atmosphere, but I found them a cruel disappointment. The meal was a kind of mash with paprika and garlic. As a national dish, it doesn't match the complexity of a Valencian paella, or Asturian beans, or Galician stew, to give just three examples. These Bettones, I thought to myself, they were some unfortunate louse-ridden devils.

The palaeontologist asked the owner of the Tsunami bar, since that was what the establishment was called (we didn't dare ask why), to sling a couple of fried eggs on top of our mash, but the man refused. He said he'd serve them to us for a next course.

In any case, thanks to the appetite awakened by the excitement and the walk, we ate heartily, while between mouthfuls, Arsuaga asked me if I remembered the second-grade levers. The palaeontologist is a compulsive teacher, and I, in turn, am an insatiable student, but from time to time I do get tired of learning, as anybody would.

"I don't get it."

"It's the nut-cracking lever. Our jaws, if you pay attention, work like that, like a nutcracker." He opened and closed his mouth exaggeratedly, gurning to make his point.

"I'll think about it," I said before changing the subject.

In the car again, heading back to Madrid, revived by the fried eggs and the coffee, I asked the palaeontologist to return to the subject of the "meddler god".

"Ah, yes," he said. "In the old religions, the relationship of humans to gods is one of pure respect, pure reverence. God asks that we make sacrifices to him, pay our respects. Full stop. But the moment comes when God starts becoming interested in what we humans do to one other. This is a god that's controlling us all the time, watching us. He's preoccupied by social conduct. In short, he's a prosocial god."

"Prosocial from the point of view of the society where he appears, I presume."

The palaeontologist pretended to be listening to me, but just kept on going.

"Not long ago, there was a paper in a scientific journal that deployed scientific methods to consider the subject. The authors asked themselves whether all those societies that developed meddler or prosocial gods had anything in common."

"And?"

"It's very interesting. First they came up with a scale by which to measure the level of complexity of these societies."

"Based on what criteria?"

"Lots of different ones. I mean, the first is population size. One million inhabitants is kind of the minimum. After that, it's about whether they had a postal system, for example, public administration, a professional army, paved roads, waterways, et cetera. Are you following all this?"

"I do follow, but don't look at me — keep your eyes on the road or let me drive."

"Do you feel unsafe?"

"No, no, you're a great driver. It's just you're looking at me too much."

"You fascinate me."

"OK — well, thanks. So we already have a certain number of conditions that define a complex society."

"Exactly. And this index of complexity goes from zero to ten. Well, it turns out that every society with a meddler god rates at least six on this index. Consistently so. Meddler gods only appear in complex societies."

"Does that mean these meddler gods create complex societies?"

"No, the other way around: when societies reach a certain level of complexity, they necessitate the appearance of a meddler god. Society is cause; God is effect."

"Right."

"But now comes the best bit. The authors of this paper look back through history, and observe a temporal delay between the onset of this complexity and the arrival of the meddler god."

"How long does it take for it to show up?"

"Several centuries."

"The prosocial god is manifested when the society has been developing in complexity for several centuries?"

"Well, before that happens, the societies have large numbers of rituals that are religious in character — they already have prophets, et cetera. With the arrival of this god, the societies become more cohesive, because the prosocial god promotes and favours conduct that's social, and punishes conduct that's antisocial."

"So it has a practical function?"

"Sure."

"And it's a god that subscribes to the dominant currents of the society in which it appears."

"Yes."

"So he's a homophobe, if it's a culture that persecutes homosexuals?"

"Indeed."

"And he orders the removal of little girls' clitorises, if that's what's in fashion in the complex society in which he appears?"

"Makes sense."

"And it's a patriarchal, machista god when the structure of

the society that calls on him is patriarchal and machista?"

"Of course. But it sounds like you're mocking me, as if I was actually saying this god exists. I'm only trying to explain to you that religious phenomena can be tackled through the experimental sciences."

"You're right — sorry."

"Strike you as interesting, or not?"

"It's interesting, but it's also a bit scary, because this god, as far as I can see, always sanctions the dominant ideology."

"Complexity is no guarantee of goodness, nor even of justice. But here's a historical example you'll find remarkable: before Columbus's arrival in the Americas, they didn't have a single meddler god. Not one. Why? Because none of the societies there got as high as six on the complexity index. To be precise, there was only one society that qualified as complex: that of the Incas. But what happens to the Incas? What happens is the Spanish arrive, and because of the delay we touched on before, at that point the Incas are still lacking a prosocial god. They aren't very far off at all — maybe a hundred years — because they've already got all the things necessary for this kind of god to appear. Even a priestly class. The prosocial god cannot appear while everyone still has their own individual beliefs. Beliefs need to be in place that are collective, regulated, fixed, universal, and organised. When all of that's there, this god shows up."

"Called something different in each place, of course."

"Yeah, but he's the same everywhere. Anyway, the Incas didn't have time to come up with their own god, even though the necessary conditions were in place, because the Spanish came along with theirs."

"But today's Western societies are complex, and yet they're secular."

"Well, that's if they really are secular … Maybe we've stopped needing God because we've got a penal code. The question is whether the UN and all the other international bodies that have substituted the meddler god are strong enough to ensure cohesion in these secular societies. The experiment of godless societies is a very recent one. We still don't know how it's going to turn out."

"So long as the gods don't change, nothing will have changed," I said, quoting Sánchez Ferlosio.

"From a speculative point of view, it's pretty entertaining, right?"

"Honestly, yes."

"I'm an atheist," added the palaeontologist, "but I don't enjoy getting involved in other people's beliefs. I don't feel that religions need to be done away with in order for things to get better. What's good about the meddler god, the prosocial god, is that he has just the same value to atheists as he does to intelligent believers. Or complex believers, we could say. The atheist would say that God is the cultural product of complex societies. A construction, in any case — like paved roads."

"And what would a believer say?"

"He or she would say it's inevitable that humankind will end up finding God, through dynamics that are part and parcel of history. God *is* there, and there's simply no way that mankind will fail to find him, if mankind attains a certain level of complexity."

"Isn't there a bit of historical determinism in all that?"

"The fact is, history operates according to certain rules: it advances in accordance with certain repeating patterns. Mark Twain said that history doesn't repeat itself, but it often rhymes. That's a literary way of putting it."

"But if individual life is the product of chance, how is it possible that collective life should be the product, as it were, of planning?"

"What do you mean, individual life is the product of chance?"

"Well, I don't know when I'm going to die, for example."

"You don't, but the insurance companies do. The individual has very little importance. I don't know what's going to become of this particular ant, but I can give a detailed account of the evolution of the ant nest. History is not a succession of merely juxtaposed events."

"So history has meaning, then? It has a direction?"

"It has patterns. And I'm not going to drop you home today, because I need to be somewhere. Jump out here and grab a taxi, or take the Metro, whichever you want."

I realised we'd just entered Madrid, but I didn't know if he'd invited me to get out of his Nissan or thrown me out. I'd learnt that the palaeontologist experiences sudden bursts of sadness that he sometimes conceals beneath an ironic demeanour, and sometimes beneath passing bad moods. I think the idea that life is absurd bothers him.

That night, at 3.00 a.m., I woke in a sweat, in a panic attack. My dreams had been visited by the two souls whom we hadn't got out of purgatory. Despite the ungodly hour, I went into the living room and messaged the palaeontologist on WhatsApp: "I can't sleep for thinking about the souls from purgatory."

I was surprised to receive a reply at once.

"Me neither. We have to go back."

EIGHT

There's no watchmaker

"Nowadays," said Arsuaga, "the dog is the king of the home, although a lot of people have them castrated. It's the only downside of being a domestic animal."

"But being castrated without knowing you've been castrated, that must be amazing, mustn't it?" I said.

It was a Saturday in late April, and the palaeontologist had asked me to meet him at noon at one of the Madrid Fairs, where there was a gathering of all kinds of pets accompanied by their owners. The dogs were the stars of the show, of course, but we also saw parrots, cats, reptiles, chinchillas, rabbits ... There was a coming and going like on Noah's Ark just before the doors closed and the flood began. People and creatures moved from side to side in search of the most comfortable spot for the crossing. Cries of panic, of warning, or of delight from different animal species were interwoven with those from the human beings, and rose up to the cavernous ceiling of the hall, bouncing off it and hurtling back down onto our heads in a shower of decibels. It wasn't easy to make yourself heard.

"What was that?" said the palaeontologist, a little more loudly.

"I'm saying that being castrated but not being aware of it must be really great."

This time I'd yelled so loudly that a lady who was standing nearby with a Pekinese in her arms looked at me with the same curiosity that the rest of the visitors were showing for other people's pets.

"Why?" said Arsuaga, not noticing the woman.

"Man, it would just be such a weight off your back. Buñuel said in his memoirs that one of the things he was most grateful for as he got older was a decrease in sexual desire."

"Oh?"

"He said that when he was a young man, whenever he arrived for a shoot in a new city, the first thing he had to arrange was who he was going to screw that night, which put him under considerable stress."

"I didn't know that about Buñuel, but, in any case, castration isn't natural."

There are plenty of natural things that make us suffer, I replied, from my own experience, to myself.

We crossed the hall, running into all types of bipeds, quadrupeds, winged creatures, mammals, egg-layers … The only creatures here with neither a pet nor an owner were us. I feared we must have looked quite odd.

"If anyone asks us what we're doing here," I suggested to Arsuaga, "we'll say I'm your pet."

The palaeontologist was preoccupied with looking for a door, which at last he found — it opened into another huge hall in which there were only dogs. The Tower of Babel that preceded it was reduced to a single language, that of barking, though still with an astonishing variety. There were dogs of all sizes, in all colours, of all breeds, of every social class.

"They must get confused with all this background noise, because dogs have really good hearing, don't they?" I said.

"Their hearing is good, but it's smell that's really their thing."

"It's their sense of smell that's predominant?"

"It isn't that it's predominant, it's that their brain is olfactory. Their mind — and that which we call the mind is the inner representation of the outer world — is olfactory. With certain exceptions, this is the case with all mammals. For them, the world is chemistry, pure chemistry. Molecules. We, on the other hand, like all other primates, create a representation of the world in the form of images. Literally, we imagine."

(We *imagine*, literally. That's so great! I made a note.)

I shut my eyes in an attempt to give shape to the world by opening my nostrils as much as I could. But I'm olfactorily blind: I was unable to reconstruct the space, despite the variety of smells my pituitary was capable of recognising.

"Sight," I said, "is the most invasive organ. And the most deceptive."

"The point here, the thing you need to take away, is the idea that our brains are visual," he said. "If a human goes blind, his or her brain doesn't change; it remains visual. Make a note of that."

I made a note: if you lose your sight, your brain, despite its plasticity, continues to be visual. Which means you're fucked.

"But sight," I insisted, "is more deceptive than smell, isn't it?"

"It's as though smells are more real. This takes a bit of thought. There is something fantastic about dogs, which is that they're the most human of all animals. More human than chimpanzees, because we've created dogs in our own image. We are their god."

"They were also the first animals we domesticated, weren't they?"

"Yes, they've been at our sides since prehistory. We are their god, and in fact that's how they see us. They do things that chimps don't do. They communicate with us, for starters. We've taught them to talk. The wolf, which is the ancestor of all known classes of dogs, doesn't bark — it communicates."

"It's true, dogs are a part of the family," I said, remembering an old documentary. "They dream of occupying our place."

"And when they try to do that, we have them put down. Those ones never reach adulthood, because they mustn't dare oppose the master."

"They may lose the battle, but they do try," I insisted.

"If they've been well bred, they won't even try. The whole thing about domestication, as you're going to see, is a big subject. We human beings are a self-domesticated species."

The palaeontologist stopped. He looked around with an expression that was somewhere between amazement and satisfaction, as if we were the gods of that whole assortment of dogs that seemed to be the extension of their masters, to whom they remained connected by the umbilical cord of a leash. There were some that walked very upright, defiantly, and others that stuck to their owner's' legs as though trying to merge with them. We discovered, at one end of the hall, a section full of special tables on which they were getting all spruced up, presumably for their beauty contests. They allowed themselves to be handled the way we do at a hairdresser's. There were also, here and there, small commercial islands popping up, on which you could find everything you could imagine to make your pet happy: food, treats, toys, collars, leads, beds, pillows …

"Well," said the palaeontologist, "we're here because it's the only way we're going to gain an understanding of evolution and Darwinism."

This line of reasoning seemed to me to be chasing its own tail, like a mastiff that we saw spinning crazily round and round on itself, but I said nothing. At that moment, we heard the sound of applause behind us. We turned and saw, a few metres away, a kind of farmyard where a particularly hairy dog was doing a demonstration of skill. Its owner was throwing plastic plates into the air, which the animal caught in mid-flight and returned. When it had retrieved seven or eight, it leapt into the man's arms, from where it acknowledged its audience. It looked pleased.

"He's really enjoying having us watching!" I exclaimed out loud.

"Do you mean he has vanity?"

"That's what it looks like."

"I'm really not sure about that. What I do know is that if his god, which is man, is happy, then he's happy, too. That's his prize: that his god is happy."

"And his god," I pointed out, "is that chubby man throwing his plates."

"Indeed."

"How many breeds of dog are there?"

"I don't know, but more and more all the time. Most of them are very recent, from the twentieth century. Before this, there were only the major branches; then they started to refine down from there."

"Refine down?"

"Yes, they chose local breeds and started making improvements."

At that moment, the palaeontologist turned his gaze on a dog that was, literally speaking, a wolf. Not a wolfhound, but an actual wolf. It was quite a shock to see it.

"Look," said Arsuaga, "the dog there that looks like a wolf, it's a Czech–Hungarian breed. Let's go and ask about it."

The owner, a lad of about twenty, told us it was Czechoslovakian.

"From which region?" said Arsuaga.

"I'm not sure," said the young man.

"He doesn't bite, does he?"

"Depends. If he doesn't like the look of you, he'll bare his teeth."

The animal was standing with his tail between his legs, clinging to his owner's thigh. From time to time, he looked up at us. He knew we were talking about him, or that was the impression I got.

"Has he ever tried to take your place as the alpha?" said Arsuaga.

"Well, not me, he hasn't," said the young man, "but he doesn't trust people he doesn't know. Me and my partner, he respects, but we always need to make it clear who's in charge. He's a part of our pack, as if he was a wolf, but he needs to be continually reminded that he'll always be behind you."

"That you're in charge?" asked Arsuaga.

"Yes," said the lad.

"And how's he finding it being here today?"

"He's nervous — you can see that, with his tail between his legs. There's a lot of people around, a lot of dogs, and he gets a bit overwhelmed by that."

"Is his hearing very good?" I asked.

"And his sense of smell, his smell particularly. They actually use some dogs of this breed as truffle hounds. This is a young one — he's nine months old. He weighs twenty-five kilos now, and he'll grow to forty-five."

"And does he bark?"

"He does, like all dogs, but when he's on his own, he howls."

"He howls to bring the pack together," Arsuaga explained.

"Right," said the owner. "When you hear him, he's just like a wolf. The Czechs created them for the army. They were after a German shepherd with greater endurance for physical work. They crossed them with a wolf, and got this breed. But it was a failure for the soldiers, because they're less docile than German shepherds. But as it happened, the man who created the breed liked them, and so he kept going with them. They're very recent, from 1955. There are quite a lot of them in Spain, but many end up abandoned because they're complicated dogs. If you leave them on their own, they'll destroy your house. They get very nervous when they haven't got their owners around. They miss the pack. They're a difficult animal. You've really got to give them some thought."

"He's a wolf through and through."

And we continued on our way, passing between dogs of different aesthetics, different cultures, between gentlemanly little dogs, with their hair in topknots, and proletarian dogs, with their hair matted, between water-dogs and anorexic greyhounds, between dogs that looked like their owners and owners who looked like their dogs.

We continued on our way, as I was saying.

And our way was back to Paley's watch.

"Paley," said Arsuaga, "was that philosopher and theologian from the eighteenth century who tried to prove the existence of God with the analogy of a watch and the workings of the world. Remember?"

"I remember," I said. "He said that if you found a stone in the middle of the countryside, you'd think it had always been there, it was a part of nature. But that if you found a watch, you'd think somebody must have left it there, because a watch can't create itself. So just as a watch required a creator, the universe, which is more complex, must have had a watchmaker: God."

"Right. And I said to you that all Darwin's theories are based on proving that the watch made itself. In other words, that nature doesn't require intelligent design. OK?"

"OK."

"And this," he continued, "was the great problem Darwin ran up against. An eye, to look at, couldn't have created itself, purely because of the randomness in the combination of its different parts. There had to be some intention, something deliberate, for a system of such complexity to come about. Darwin believed in evolution. He believed that species evolved and modified over time, without the need for any 'watchmaker,' but he couldn't find a way to explain it — the mechanism evaded him, he could find no reason why. In science, if you don't have an explanation, you might as well have nothing. You can observe the sun rising in the morning and setting at night, but if you can't provide a reason, then all you've got is the observation."

"So, Darwin," I said, trying to understand, "needed to demonstrate that species, in nature, evolved without there being a need for any purpose to exist behind that evolution."

"Exactly. How can the perfection we see in living creatures be attained without the pre-existence of some design?"

"Well, how?"

"Darwin spent years studying the domestication of animals. He sensed that there was something in common between the production of domestic breeds of dogs and evolution, but he didn't work out what it was until he came up with the idea of 'unconscious selection'. No one's ever recognised the true importance of this discovery."

"Well, you have."

"I wrote a short book on the subject, because it seems crucial to me. Darwin discovered that nobody in antiquity tried, as people do nowadays, to create, I don't know, a breed of horses to compete at the racecourse, or a breed of cows that would give you lots of milk. Or a guard dog, or a homing pigeon. These things that we do nowadays are called 'conscious selection'. All of the breeds of dog on show here today are the result of conscious selection. But in antiquity, people simply worked with whichever animal was most useful to them, and it never even occurred to them to create a particular breed. If a kind of sheep gave you a lot of wool, you'd breed more of it, and the one that didn't, you'd put in the pot. If there was a particularly large kind of corn growing, you'd sow more of it. In other words, there isn't that much of a leap between conscious selection, which is what we do nowadays — aimed at upgrading the breed of this or that creature — and the kind of unconscious selection that occurs in nature."

"It doesn't seem correct to me," I pointed out, "to say that a farmer or a rancher's decision to keep the sheep that gave the most wool or the kind of maize that produced the most corn was unintentional."

"But there wasn't a *conscious* intention. The Jerez vineyard dog, for example, turned out to be ideal for keeping rats out of your wine cellars, because it was small, because it could get into all the nooks and crannies. There weren't any show days for vineyard dogs. The ones that were best at performing a certain function were allowed to reproduce, and that's all there was to it. Domestication, when all is said and done, is the control of reproduction. Write this down: domestication consists of the control of reproduction. Yes or no?"

"Yes."

"What do we mean by a domestic species? That you control its reproduction. You decide which ones get to reproduce. You *select* which ones reproduce, and which ones don't."

"And, to a great extent, that selection was unconscious."

"In antiquity, yes. Well, this is Darwin's idea when it comes to nature: unconscious selection. There's no watchmaker, no planning, no objective. No direction, no intention. The creatures that adapt best to the niche in which they find themselves survive, and they get to reproduce. All the perfection, all the beauty we see in nature is determined by death. What lies behind the harmony you see in the countryside is the Reaper with his scythe."

"And those that perish are the ones Bataille calls the 'accursed share'. He wrote a book with that title."

"Call it whatever you want. A cheetah can run 90 kilometres an hour. If one of them, for whatever reason, only manages 85 kilometres an hour, it's dead. Being a sub-90-kilometres-an-hour cheetah means being dead."

"You do get examples of perfection, then, in each individual case."

"Darwin's very insistent on this point: there's no such thing as perfection in general, only perfection in specific circumstances. Machines and living beings alike can only be judged according to their efficacy within their given sphere, in the place they occupy in the economy, in their market."

"In their market?"

"At that time, Darwin was reading Malthus, who is the

founder of demographics, and who had written the book in which he said it's no good helping poor families because that will only mean them having more children, leading to an increase in mortality rates. Which means that if no brake is put on the population, it will snowball, while resources will grow at a slower rate. If there isn't any control, in short, you get scarcity, you get people in destitution, and you get conflict. When Darwin read this, he said: 'Right, many more wolves are born than can survive.'"

"Many more wolves?"

"And deer and robins and rabbits, anything you can think of. There's a concept in ecology called 'carrying capacity'. Essentially, each cow, or, I don't know, each auroch, let's say, needs five hectares of pastureland. There isn't enough to feed any more than that — you can't fit more cows, more deer, more lions. The first port of call when designing a nature reserve is to ask oneself: how many goats could we fit here? And, say, five thousand can fit. Write it down: carrying capacity."

"But in nature, this regulates itself."

"Yes, it's regulated by death. The law of the jungle. Cheetahs that can run more than 90 kilometres an hour get to survive. There you have it, that's the reasoning. The immense majority of goats that are born are sure to perish. Natural selection can be unforgiving."

"The accursed share," I insisted.

"Call it what you like," he repeated. "Bats are simply perfect at being bats."

"But they make for terrible moles."

"Now you're getting the hang of it. This is how Darwin, reading Malthus, came up with the key. Unconscious selection: they compete amongst themselves. He realised that, although everything in nature appears to be alive, in reality almost everything's dead, because of natural selection."

"So, no watchmaker, then."

"There's competition, selection, and a very slim survival rate. This goes for all species, humans included. You and your wife could have something like sixteen children, but out in nature, only two would survive."

"Yeah, that's crazy."

"This is why people so often get Darwin wrong, because his discovery paved the way for so many different things. And because he took inspiration from demographics and economics, lots of people have used his work to justify the status quo. 'Darwin says so,' they argue."

"Did Darwin read everything that happened to fall into his hands?"

"Everything. Look, there's no written evidence of this, but many of us believe that Darwin's greatest single influence was Adam Smith. Smith believes in the 'invisible hand' of the market. He says it functions of its own accord, that there's no need for any intervention. That's the basis of liberalism. The invisible hand of economics regulates everything, and leads to progress amongst nations. If left to its own devices, the economy will give rise to different specialisations: carpenters appear, and bakers, and bricklayers … A variety of different jobs will come about without any planning, because people, according to their aptitudes, will occupy a place in this complex system that is society, and society will progress just as nature progresses. There's an economy of nature: species adapt in order to get better at a certain function, to occupy a niche. Darwin never said anything about having read Smith, but I'm sure he did so in the autumn of 1838."

"You mentioned progress. But what do we understand progress to be?"

"With very simple forms as its starting point, life has unfolded, has undergone a process of perfection."

"Complexity as a kind of progress?"

"Yes, on the one hand. But on the other, in Darwin's time, there was a general feeling of optimism around. In the Victorian era, society was believed to be progressing in all different spheres, and this progress was thought to be unstoppable. There was more wealth, more comfort, improved health, increased happiness. The concept of progress was etched into English society at that time."

"And what happens to the disadvantaged classes?"

"Progress would also reach them. There's this overflowing euphoria. Then, come the second part of the Industrial Revolution, everything starts getting complicated. The growth of factories, mining, all the gruelling jobs ... We see the rise of the urban proletariat ... But in Darwin's time, there was still this transition, with the rural poor — who'd been lorded over by a tremendously wealthy aristocracy — becoming urban, and living better lives than they had in the countryside. There's the growth of cities ... Plus, the English are gaining an empire."

"There's a feeling of power."

"There's a sensation of unstoppable progress, and this has an influence on Darwin, because he is a Victorian. In any case, Adam Smith put forward an economic model that also provided an explanation for the history of life on earth."

"And was Darwin a social Darwinist? Did he think it was a good idea to apply the laws he'd discovered in nature to relationships between humans?"

"No, Darwin was a very good person. He was against

slavery, for example. The problem isn't the transfer of economic theory to natural law, but the other way around: when natural law gets applied to economic theory."

At this moment, we stopped at a stand where dogs of the haute bourgeoisie were being shown. The woman judging them watched how they walked with their owners, calculated their heights, assessed the position of their legs, evaluated the shape of their ears, the length of their backs …

"Oh, look," said the palaeontologist, "they're doing morphological examinations of these dogs. Look how worried the owners are! It's like they're having to do one of those terrible tests to become a postal worker."

"Or to become a notary," I said.

"Or a lawyer."

"Or a palaeontology professor," I joked.

"That'll do," he said. "Have you made a note about how important the observation of domestic animals was for Darwin?"

"I think so."

"In that case, let me buy you a beer, and then I'll have to dash. I'm going to Communion."

"Holy Communion?"

"Didn't you notice I'm wearing a suit jacket?"

NINE

Super teddy bear

June marked the completion of a year since my first meeting with the palaeontologist — a year in which our cholesterol hadn't gone up, nor had our blood pressure, and we hadn't had a roof tile fall on our heads. Compared to the progress of the world, our lives were proceeding with no great noteworthy shocks. The partnership, in short, was working. I called him to say we should celebrate the occasion, and he agreed.

"I'm taking you to a toyshop," he added.

After I hung up, I was a little concerned. Was he thinking of buying me a teddy bear as an anniversary present? Had he started to sense my profound Neanderthalness? If that was the case, what gift ought I to buy him?

What can a Neanderthal offer a Sapiens? He told me to meet him in a shop on Madrid's Calle del Arenal at seven o'clock on a Saturday evening. Calle del Arenal, which is pedestrianised, joins the Puerta del Sol with the Plaza de la Ópera, two of the city's nerve centres. The artery was teeming with people, the way a petri dish teems with microorganisms in the lab. I arrived half an hour early, as I usually do, to inspect the surroundings, and I had a peek into the establishment, which was, indeed, a toyshop, whose aesthetic evoked an English establishment in the nineteen twenties. The window displayed dozens of hyperrealist babies, but also stuffed animals and even a doll's house.

Dolls' houses drive me wild. The one in the window was two storeys high with an attic, and it was half-open, showing its innards: the living room, the kitchen, the bathrooms, the bedrooms … In the living room there was a group of older people drinking tea. In one of the bedrooms, a girl, who reminded me of Carroll's Alice, was looking at herself in an oval mirror, one of those big cheval-glass types. In the attic, a butler and a cook were conversing while seated on the edge of a high bed. It looked like a peaceful world, perhaps too much so. The only addition I'd have made, on the lower floor, in the gap under the staircase, was a hanged man swinging from a beam.

Shortly afterwards, I began to wonder whether Arsuaga really had asked to meet me there, or if it had just been a dream. My suspicion grew when the agreed time came around and he hadn't arrived. I went into a nearby bar from where I'd be able to watch the entrance of the shop, and ordered a coffee to kill time and reflect upon my mental state. Around a quarter past seven, when I was about to leave, I saw him arrive, hurrying slightly, making his way through the crowd.

"Sorry, sorry!" he said. "I've just driven in from the mountains. I was hiking, and I got caught up in traffic."

I asked him what we were doing there.

He turned and gestured to the throng, and exclaimed: "Look, all this energy!"

I hate energy, I hate euphoria, I hate large crowds, but I feigned enthusiasm at that Saturday-afternoon spectacle in the middle of one of the great European metropolises.

"OK, now I've seen the energy," I said, after a few moments. "Now what? What are we going to do in a toyshop?"

"There are lessons to be learned everywhere," said the palaeontologist, with a somewhat condescending smile.

The mountain air had acted on him like a line of coke. In addition to this, he'd had a haircut that made him look like a teenager. He was wearing a T-shirt and jeans. He seemed particularly thin. For a moment, I found him rather loathsome, if I'm honest.

"This ebullience," he said, not moving from where he was, "is all somatic. It's to do with the body, even though every one of these people carries a fixed set of genes inside them. Have we talked about this already — germlines and somatic lines?"

"Doesn't ring a bell."

"The body is the vehicle for the genes. There are those who say that, if it came down to it, the genes would do without the body altogether if it were to their benefit, because they are selfish. It's one way of looking at it. In the chicken/egg dichotomy: we choose the egg, but there's an aphorism according to which the chicken is nothing but the instrument used by the egg to perpetuate itself."

"The chicken is the shell."

"Something like that. All of these people will die, and so will you and I, but our genes will carry on through the ages. That's precisely what they've been doing since time began."

I imagined that whole crowd dead, including the hundreds of teenagers who were going in and out of the countless bars, and the idea seemed like total carnage.

"Let's have a look at that koala," Arsuaga said, going over to a two-metre teddy bear next to the Church of San Ginés, with which children were having their pictures taken.

"And the toyshop?"

"Later. We've got time."

We made our way between the bodies until we reached our goal.

"This is a super teddy," he said, pointing at the monster. "The koala is, in and of itself, a teddy bear animal. We love stuffed animals because they bring out our tender side. Our genes manipulate us to bring out a protective impulse."

"Well, this one is actually a bit scary," I said, contemplating its size.

The palaeontologist just went on: "A protective impulse similar to that which we feel toward children of our own species. We don't consider children a threat, do we? They aren't part of the system: they don't play the social game in which we adults are involved; they don't compete. This is essential if they are to play on our unconscious emotional mechanisms, the inherited, genetic ones — our biology."

"And that," I ventured, "is why horror movies with children in them are always twice as terrifying: because the threat's coming from someplace it shouldn't."

"A devil child is the most terrible thing there is. But what's so interesting about teddy bears? What makes the koala so adorable?"

The koala's owners, a Latin American couple, and the people who were standing in line to take photos of their children, were starting to give us curious looks. What were two older gentlemen doing, planted in front of this creature, in animated conversation — one of them taking notes of what the other was saying?

"I'm afraid our presence might be making them a little uncomfortable," I said.

"Forget about other people's discomfort," Arsuaga scolded.

"You spend your life worrying about what people will say. To start with, the koala is cuddly and soft, and its fur isn't spiky; it's strokable. Do you see?"

"I do."

"It's one great big ball. Now, we're going to break down what it is that makes children adorable, and the characteristics they share with teddy bears. Firstly, the cuddliness. They have to be like a round ball, with barely any neck. The head is a sphere. They don't have sharp teeth or claws."

"The koala's got claws."

"But they're retracted. Whereas a wolf in the wild has sharp teeth. Look at the koala's face. Big eyes, short snout, and a round forehead. All the things you see in a child's face. And what about the way children move? They're uncoordinated, always on the verge of falling over. That lack of coordination, it's fundamental in awakening our tender side. Plus the short arms and short legs. Add all these things together, articulate them correctly, and you've got a machine for bringing out the tenderness of others. The genes responsible for producing these features are working on your conduct. They are manipulating you, and they aren't even yours."

"Not even the same species as me," I added, "because a puppy awakens the same emotions."

"Exactly. This is what we're here to talk about today, because the last time we met was at the dog show, remember?"

"Yes."

"Why do we like dogs, why do we find wolves threatening, and why have we invented pets with familiar features?"

"I think I'm starting to see."

"And now there's another interesting word I want to mention, another key concept, which is the *super-stimulant*.

Every kind of manipulation, from totalitarianism to sexual persuasion, to advertising, uses this. Children are already pretty cute as it is, but if you make a super-child, then you have fabricated a super-stimulant. If you exaggerate their features, they draw the attention more."

"A super-koala is a koala that's been modified to provoke more tenderness than a regular koala," I ventured.

"Indeed. It's an exaggerated version of a koala. Look at how trusting the children are in letting it hug them. And how, in spite of its size, they hug it without any kind of fear."

"You're right, but maybe we ought to go to the toyshop — they might close soon," I encouraged him, troubled by the curiosity we were prompting among the ring of bystanders.

"Well," said Arsuaga, ignoring my suggestion, "this applies to everything."

"Such as?"

"A cake made with too much refined sugar and lots of fat."

"Those calorie bombs ..."

"What are those cakes? Super-stimulants. We like sugary fruits. We are programmed to eat blackberries because they've got glucose in them, and we like animal fats because they give us energy. As well as needing proteins, which are the building blocks of the body, we need energy, and we get energy from sugars and fats. To obtain fats out in the wild, you have to hunt a mammoth, and that takes lots of time and effort. A single cake has all the fat you would get from a mammoth."

"And to obtain the sugar you get in a slice of cake?"

"You would need to eat every single blueberry in the Sistema Central to match the amount of energy you can get from a single slice of cake. How, then, to resist the super-stimulant that is a cake?"

"With willpower," I replied, though this was clearly absurd.

"Biological super-stimulants," he continued, "are common throughout the species, so that, if you want to sell something, you already know which button you need to press. And now, yes, let's go to the toyshop before it shuts."

Once we were inside the shop, and having explained to the manager that we weren't a pair of dirty old men, but a palaeontologist and his student, we stood amazed at a collection of latex dolls that provided a perfect imitation of the texture of a baby's flesh. They awoke not only tenderness but also cannibal instincts, as they looked quite ready for the oven. I asked the palaeontologist if the phrase "I could eat you all up", which is so often used to refer to children, did deep down express a literal desire.

"My mother," says Arsuaga, "has a story about how, not long after my older brother was born, she was served suckling pig and she said, 'I can't eat that.' Maybe it reminded her of a desire to eat the baby boy — who knows? — but the truth is that babies are made to be eaten."

"Talking about the cannibal thing, I'm reminded that at home we used to have a pair of hamsters, and at one point the female had a litter. And I remember, one day the mother seemed to be doing something strange, and I approached the cage. Turns out she was eating one of the babies. She had taken it like this, between her front legs, the way a squirrel takes an acorn, and had started with its head. I still get chills; I'll never forget it."

"In my house," Arsuaga said, "it was the boys, my kids."

"Who ate the hamster?"

"No! Who came upon the mother hamster eating her offspring in the master bedroom."

"How appalling!"

111

"The genes, it's the genes — it's nothing personal. In reality, she wasn't eating them, she was recycling them. When a female hamster gives birth inside a cage, she feels she's in an unsafe situation, and the best that she can do therefore is to recycle the energy of her offspring. The litter is doomed."

"Right."

"But anyway," he said, returning to the hyper-realistic dolls, "we see here the characteristics that make children so sweet and lovable. The same things we were saying about a koala bear: disproportionately big head, big eyes, chubby cheeks, no sharp corners, rounded forehead, flat nose, almost furry on top, head barely peeking out. Can you imagine a baby with an aquiline nose?"

"No."

"And the lips, those little chops ... Plus, they've either got no teeth, or very tiny ones. Everything's all soft and bouncy: the little tummy, the chubby thighs ... And its clumsiness — again, clumsiness really makes us go gooey inside. What's the baby saying to us with all of this?"

"What?"

"*I'm not here to compete with you.* A baby is a survival machine. It's programmed to reach adulthood. Make a note of this: we could use all the features we've just seen together or apart. Once you've got a list of features, you say to yourself: I'm going to amplify them all, or just one of them, maybe two, et cetera. And away you go, manipulating people. Let's go into the next room — that's where the teddies are."

"The curious thing," I insisted standing now in front of the teddy bear display, "is that it isn't just the young of our own species that induce tenderness and an eagerness to protect them, but animals, too. And the same thing happens to animals with us. That's what's behind those cases of feral children who get raised by an animal in the wild."

"That is exactly the point. It's something you see in all mammals, bar none. The same infantile features recur across them all. Hence the stories on TV about a lioness adopting a newborn from another species. The lion is no zoologist, so it doesn't know, but this newborn's appearance awakens her protective instinct. The lion isn't in control of this instinct. All mammals are alike in this."

"Of course," I said, "while a baby earthworm, on the other hand, doesn't provoke any feeling of solidarity in us at all."

"Look at that husky," Arsuaga says, pointing out a nearby husky puppy, "it's saying: 'Adopt me.' It's manipulating you to make you want to adopt it."

"It's true!" I exclaimed, amazed.

"If you like it, I'll get it for you."

"What?!"

"I'm joking — don't get your knickers in a twist. The majority of dog owners will swear to you that they didn't choose the animal, it chose them."

"How so?"

"You go into a pet shop, and all of the dogs start rolling around and playing cute, as a way of seducing you. They compete with one another to make you like them. And you go away with the one that's touched you most deeply."

"And so they choose us."

"Indeed. All these teddy bears, if you look, have something in common. What?"

"What?"

"Their needy attitude. They all look up at you, not so much asking for help as for affection. Yes or no?"

"Yes, but that bird", I add, gesturing toward a stuffed raven, "doesn't look all that tender to me."

"With birds, they do what they can. Make the beak a little rounder, for example. Me, I like the octopus over there. Look how cute it is."

"But the octopus is an alien."

"The octopus, in spite of its morphology and the fact it's related to clams and oysters, has developed several features very similar to our own."

"I've heard that."

"For starters, this animal has a mind. A mind is what machines lack. It means you've got an inner representation of the external world. A replica. This is what we know of the world: a replica that we have inside our heads."

"The head's a bit like Plato's cave: all you can detect is an echo of reality."

"That's one way of looking at it. What's for sure is that no machine has a mind. This is why computers win at chess, but they lose at Parchís."

"It's strange how an octopus can look so like us while we're so different in shape."

"This is called adaptive convergence. I'll explain it to you in detail some other time. But think, for example, of Hernán Cortés and Moctezuma. All the Aztec institutions were things the Spanish conquistador recognised: they had priests, schools, books, churches, monarchs, soldiers, generals … Cortés could understand their society perfectly well, in spite of the fact there had been no contact with it for years. The humans who arrived in the Americas fifteen thousand years before the conquistadors were mammoth hunters, and now these people were writing books, same as us. So, what can we deduce from that?"

"What?"

"That there are cultural convergences determined by the nature of our mind. There are paths that repeat. And the octopus is a good example of this. We separated from the molluscs millions of years ago, but we have converged mentally with the octopus, which has an eye that seems to be looking at you."

"With an expression that's human."

"Almost. But now that the subject of convergences has come up, I'll confess that there is a world, the world of Pokémon, that I am about to get into."

"Did you just say you were going to get into Pokémon?"

"Yes, because from what I can make out, they're these fantastical animals, chimeras, creatures made up by combining, I don't know, a rabbit and a cat, which in real life is completely unfeasible. There's an internal logic to evolution; it isn't like anything goes. You can't have a carnivorous rabbit. You can't have a rabbit-cat. You also can't have a carnivore with horns. There's a story about the Devil appearing one day to Cuvier — he's the father of palaeontology — and saying to him: 'I'm the Devil and I'm going to eat you.' Cuvier looked him up and down, and said, 'You've got horns on your head and hooves on your feet, you can't be carnivorous.' And he turned over and went back to sleep. Because he was in bed."

"Real tough-guy, that Cuvier."

"Adaptive convergences exist because the number of possibilities is limited. And that makes for apparent coincidences between creatures, like the octopus and you, which look very different."

At this point, we were approached by the shop manager, who told us they were closing.

"What a shame!" Arsuaga said. "We've still got dozens more teddies to look at. But that octopus really is extremely well done, isn't it?"

The woman gave us a suspicious look. She clearly would not have believed we were a palaeontologist and his student. On our way to the exit, we stopped at a dolls' house identical to the one in the window, and I asked the palaeontologist: "What's missing?"

"I don't know, what's missing?"

"A hanged man in the gap under the stairs."

He looked at me.

"Are you OK?"

TEN

Two skaters

In late May, Arsuaga had published a book entitled *Life: the great history*, which was subtitled *a journey through the labyrinth of evolution*, and which I needed to digest in two stages, like a ruminant. First, I read it anxiously, without fully understanding it, and then I regurgitated it and re-chewed it, softening it with my mental juices so as not to miss out on any of its nutritional content. When I was in the middle of the process of ruminating, well into June now, I was invited to introduce the book at an event at the Espacio Fundación Telefónica. Well, more than just to introduce it: to have a public conversation with its author. Though the idea worried me, I couldn't say no, given the bonds with the palaeontologist that had already been formed.

I arrived an hour early, and ordered a gin and tonic at the bar of the Hotel de las Letras, which was right next door. Then I received a call from Arsuaga.

"Where are you?" he said.

"At the Hotel de las Letras, having a gin and tonic."

"Why?"

"Why do you think?", I said. "To settle my nerves."

The palaeontologist fell silent a few moments, during which I thought he would decide to join me, but he merely informed me that they were waiting.

The chat went well. I explained, at the outset, how the book was structured: divided into two parts, the first was devoted to the evolution of species; the second, to human evolution. The alcohol, while not turning me euphoric, had provided me with just the right amount of energy to give the gathering a conversational tone far removed from an academic register. Arsuaga played along, so that a mood was quickly established in the room — which was full to bursting — that was relaxed and friendly, and was reflected in the happy expressions on the audience's faces. We talked with the ease of two skaters who manoeuvred on the ice, our paths crossing and uncrossing, as we sketched intricate rhetorical patterns, never bumping into one another. I was surprised at how the palaeontologist had managed to reconcile academic rigour with a popularising ability that brought the book within reach of anyone prepared to make the effort that all truly beneficial reading demands. The volume returned the effort you invested in it several times over.

But there was another matter that had engrossed me, and it was the fact that, beneath the rational discourse, and subject to the rules of scientific speculation that ran through Arsuaga's writing, deep beneath the volume's surface, I was aware, however faintly, of a shiver of an existential nature. That learned man who was so sure of himself was nonetheless doubting. I asked him about this matter, on which he expounded at length, quoting Unamuno's *Tragic Sense of Life*.

As he talked, I realised what a great sense of the theatrical Arsuaga had. He was a master of oral storytelling. He knew when he had his audience, and when he was running the risk of losing them. He endeared himself by combining intellectual precision with a kind of helplessness — actual or performed — that delighted his listeners. That mixture of wisdom and people skills made me a little envious.

When the performance was over, I left him signing copies, without saying goodbye, because there was a queue at his table of at least forty or fifty people. We wouldn't see each other again till September.

ELEVEN

All children

I spent July and August at my house in Asturias, from where I sent the palaeontologist several emails to which he replied monosyllabically. He was involved with something important to do with his archaeological digs, or that was the impression he gave me. I didn't manage to establish an epistolary relationship that could diminish the feeling of rupture prompted by the summer break. I also invited him to visit me, luring him with all the guile of a spider crab, and he promised me he'd try, but he never came.

For a time, I hated him a little.

In September, no sooner were we back in Madrid than we arranged to meet in a Japanese restaurant beside the Gran Vía. I hoped the sight of raw fish might prompt him into telling me about the importance of cooking, as over the summer I'd read a really interesting book about the taming of fire and the changes the new diet had wrought on our digestive systems. But Arsuaga appeared a little stricken by the academic bureaucracy he'd run into since his return to the university, and he didn't fall into my naïve Neanderthal trap. Besides, he was about to turn sixty-five and to marry off his son. I told him that maturity always rings twice, and he asked me if I was calling him old.

"Not at all," I said hurriedly. "You've lost weight, by the way."

"Well," he said, "I've taken up running."

He'd just been at El Rastro market to buy some analogue film to record his son's wedding, because he wasn't convinced digital recordings would really last.

"Turns out analogue things are sexy. Super-8's are sexy, celluloid is sexy."

"Yeah," I conceded.

When they brought out our second dish, the palaeontologist glanced around the restaurant, which was full, and gave a smile that was either enigmatic or ironic — I couldn't tell.

"What's up?" I said.

"Have you realised just how many people there are here, and yet how calm we all are?"

"Why would we be nervous?"

"What I mean is, we're a domesticated species."

"So who's our owner?"

"First, let's look at all the markers of this domestication. What are the similarities between domesticated breeds of dogs, cows, and sheep?"

"I don't know."

"To begin with, a pronounced sociability: they're all very gregarious. That makes it possible to organise them into herds. This is the reason we've domesticated them. We aren't interested in solitary animals."

"So you can't have cat farms?"

"No."

"But cats are domestic."

"Not so much. Think about us. It's as though we had an

ancestor who existed in the wild, but we've domesticated ourselves. Otherwise, how do you explain just how peaceful it is here? If we went down to the Gran Vía, we'd see huge numbers of people, and none of them fighting. We've got a great tolerance, an enormous capacity, for grouping together, for joining in a pack with other members of our species with whom we have no blood relation, whom we don't know from Eve. Wolves, you can't put them in a room with wolves from other packs — they'd just tear one another to pieces. We'll put aside the question of the difference between herd instinct and sociability for now … But, to simplify: the domestication of the human species begins with the fact we're so social."

"And it's never been possible to transform species that weren't social into social ones?"

"Never."

"What are the main features of a domestic species?"

"Submissiveness, docility, lack of aggression."

"And how are those achieved?"

"By infantilisation. As I said to you not long ago, dogs never reach adulthood; they always remain children. If they were adults, they wouldn't be able to coexist with us, and they'd be in constant dispute with the master."

"Some do try."

"And they get punished for it. If they keep on trying, they get put down. This is how we've managed to turn the wolf into a companion animal: by allowing only the most docile to reproduce. Do you remember what I said domestication consists of?"

"Controlling reproduction."

"In essence, yes. You domesticate when you decide who gets to reproduce and who doesn't. According to what criteria?

According to which domestic animal will be useful to you: give you milk or wool, keep you company, pull your cart, or guard your home. Whatever you decide. Every domestic breed has a practical function. But there's one characteristic none of them can do without: they have to be docile. Elephants, for example, have to be docile because they are very strong. For this to be possible, it's necessary that they live together, in herds."

As he talked, Arsuaga was demolishing the rice and sushi with the tips of his chopsticks. He didn't like it, but he didn't complain; he didn't say anything. He merely dismantled it. *He's docile*, I thought.

"And what you were saying about infantilisation — that applies to humans, too?" I asked before bringing a smoked-eel roll to my mouth. "Do we ever grow into adults?"

"In effect, we human beings are at play our entire lives; we never become adults. Observe, for example, the passion that is awoken by football."

"I don't like football."

"It doesn't matter what you like. We are something else's domesticated species."

"And for a human to become an adult, what would that involve?

"For a dog, becoming an adult would consist of turning into a wolf."

"OK. And for a human?" I pressed him.

"Into a Neanderthal, turning into a Neanderthal," he said, and a terrible silence fell between us.

Into a Neanderthal, I said to myself. *That's what I am. So I'm one undomesticated guy living amidst tame creatures? And could it be that Arsuaga is a clandestine Neanderthal himself, a Neanderthal who's pretending to live having adapted to the Sapiens' rules?*

The palaeontologist had noticed the confusion that his statement had caused, and qualified it: "OK, let's go into this a little more. That was just a way of putting it."

"What you've said is that for us to become adults we'd have to transform into the species that didn't survive."

"Partly, yes. We are the domesticated species of Neanderthal."

"Could the Neanderthal not be domesticated?"

"No, it *was* domesticated. That's us. You and me. But be patient, we'll get there. For now, we are infantile in everything. Physically infantile. Our brain is smaller: it's shrunk over the course of evolution."

"And did that shrinkage mean a loss of abilities?"

"It's shrunk in the same way domestic animals' brains have shrunk. A cow's brain is smaller than that of an auroch. A dog's brain is smaller than that of a wolf."

"And our brains are smaller than ..."

"That of Cro-Magnon Man, the one from Altamira, who was in the picture twenty thousand years ago."

"But from a cognitive perspective, we're superior."

"I don't think so. Think about those bison they painted in Altamira."

"So let's see," — I was trying to get things straight in my head — "what made us human was the increase in the size of our brains. Yes or no?"

"Yes."

"But their subsequent reduction hasn't dehumanised us."

"On the contrary, it's one of the things that produced the modern-day human. What you understand as being human is a docile creature, a tame one. A savage, someone who's aggressive, isn't to your mind properly human. Remember what Jesus

Christ said: *Except ye be converted, and become as little children, ye shall not enter into the kingdom of heaven.* This, in a literal sense, is what we have done."

"But we haven't made it into the kingdom of heaven," I retort.

"Might this not be the kingdom of heaven?" he said, motioning around at the establishment.

"Well," I conceded, "we are in a good Japanese restaurant, working our way through some incomparable sushi, surrounded, it's true, by people who don't attack us, who don't try to take our food away, we're keeping up a pleasant conversation ... Yes, maybe this is the kingdom of heaven."

"What more could you ask for?" Arsuaga said, with a wry smile. "Would you rather be surrounded by a choir of angels? Do you have any interest in joining a choir of angels?"

"I prefer this," I said, gesturing toward the crab with the soft, crunchy shell.

"This is as good as it gets," he said, still rather caustic. "We really have reached the top."

"So we're like children thanks to the diminution of our brains," I said.

"Konrad Lorenz said that human beings retain their curiosity and capacity for play their entire lives. If you don't like football, just think about some of the horrifically infantile programmes on television, whose main audiences are elderly people. Lorenz said that an adult lion is an extremely serious individual; you don't mess around with adult lions. And an adult gorilla is the most serious thing there is. I've been to Rwanda, looking at old gorillas, and I can assure you they don't play at all, they don't laugh at anything. There's nothing more serious than an elderly gorilla."

"Are they undomesticable?"

"Absolutely. Do you know what neoteny is?"

"Haven't got a clue."

"The capacity for the adult organism to stay young. 'Forever young.' Who sung that again?"

"It's Dylan, I think. What are the other consequences of brain shrinkage and domestication?"

"A loss of sensory acuteness. Wolves have a better sense of smell and of hearing than dogs. When a wild species like the wolf begins to be domesticated, strange and very varied features start to appear: their ears droop, you get blotches of colour. Whereas when a domestic species goes back into the wild, it returns to its original state. If you give a wild dog long enough, it will return to being like a wolf, because natural selection will favour the most aggressive. In those conditions, only the toughest survive. Meaning that, if we were to turn feral …"

"We'd revert to being Neanderthal," I finished his sentence.

"Well, we're going to see if that exists."

"If what exists?"

"Anything that's actually like us. Look, there's a very elegant example in the shape of bonobo monkeys, which are the twin species to chimpanzees. The two are separated by the River Congo. Chimps are very aggressive, and they have a male hierarchy where the pecking order is very clear. The males dominate the females in such a way that the lowest of the males is still better off than the top female. They're very aggressive, very violent. Very territorial. But with the bonobos, it isn't the males that dominate; it's the females. And they resolve all their problems with sex. They're the hippy version of the chimpanzees."

"How'd they manage that?"

"Very simple: the female bonobos create alliances. The

weakest of the males is stronger than the strongest female. But the strongest male is weaker than multiple females who form an *alliance*. Do you follow?"

"Yes."

"But there needs to be some agent doing the selecting. The male bonobos, though strong, are quite pacific, because they're the ones who have been chosen. Who's the agent of this selection, who's made this choice?"

"The females."

"The females, who over the course of time have been doing away with any aggressive males."

"How?"

"By killing them, of course. Stopping them from reproducing. And expelling them from the group, which amounts to the same thing. As Lorenz said, a solitary chimp isn't a chimp. Either it's social, or it isn't a chimpanzee. And, of course, a human being living in isolation isn't a human being, it's something else. A human being can only exist as a part of society."

"Do you remember the José Agustín Goytisolo poem that Paco Ibáñez used to sing: 'A man alone, a woman, thus taken one by one, are like dust, they are nothing, they are nothing'."

"Exactly — you can write exactly that."

"Thanks very much."

"I'm currently working on the speech for my son's wedding — it's a gay wedding. I'm going to be master of ceremonies. It's prompted me to do some reading around on the subject of love. By this marriage, our children have improved us, their parents, they've broadened our horizons, our tolerance, they've helped us to grow. Our perspective on the world has expanded. It turns out they're going to be married on the 28th, and the 29th is

Saint Michael's Day, the day of the three archangels."

"Indian summertime, it's called in some places, or quince time," I said.

"Right, and it's called quince time because it's when the quinces ripen — the fruit that was once dedicated to Aphrodite. Venus, to the Romans. In those days, anyone getting married would go to the temple of Aphrodite and be given the gift of a quince before entering the nuptial chamber. Quince was a guarantee of love and fertility. See, haven't I got a lot of interesting stuff for my speech? I've read some surprising things. One that jumped out at me is about the relationship between parents and children, and the way children improve their parents. Parents educate their children, but then they are educated by them. I don't know who it was that said that children, in fact, give birth to their parents."

"Our children give birth to us?"

"Yes, exactly that. And I also found a quote from a modern-day English poet. I can't remember his name, but it's to do with people saying: 'I will always love you.' And he says that it's very simple to say you'll always love someone. But what about promising you will love them next Tuesday at 4.30 in the afternoon?"

"That's complicated," I confirmed.

"Complicated, yes. How the hell did we get onto this?"

"It was about how we're the domesticated species of the Neanderthal."

"Ah, yes, the bonobos. Separated from the chimpanzees, of which they are a twin species, by a river, and with a completely different biology. A social biology, not a cultural one."

"What do you mean exactly?"

"That it's in their genes, in their genetic programming. It

isn't that the male bonobo gets done over, or dominated, or suppressed; it's that he is by nature pacific and tolerant, because it's the most pacific and tolerant individuals who have been selected. And by whom? The females. A very well-known primatologist, Richard Wrangham, says that we are the bonobo of the Neanderthal."

"Being domesticated, but arseholes," I said, recalling some statements by Donald Trump.

"The two aren't mutually exclusive. Anyway, to conclude: we have all the infantile features of domestic species: this high forehead: this lack of prognathism, our being neotenic, being teddy bears. To the Neanderthal, the Sapiens must have seemed like a teddy bear. This is described very well in a novel I helped get published, *Dance of the Tiger*, by Björn Kurtén. I recommend it."

"I've read it, but I always thought it was the Sapiens who was the arsehole. That the Neanderthal screwed the Sapiens out of love, while the Sapiens screwed the Neanderthal out of self-interest."

"Don't get mixed up: the arsehole, the aggressive one, was the Neanderthal. And to them we would have seemed infantile. Neoteny consists of you resembling your ancestors without having lost their infantile features. The Sapiens resembled the Neanderthal's offspring."

"So the Sapiens acted like a Trojan horse. They went into the Neanderthal's home all innocent-seeming, but ultimately, who was it that survived?"

"Look at it however you want. The question is: who is responsible for the domestication?"

"In the case of the bonobos, we know that already: the females."

"Now we need to work out who's done the selecting in the case of humans."

"Which would be equivalent to answering the question of who our owner is."

"No, no."

"Whenever I reach a conclusion on my own, you always say no," I complain.

"It's just that you take everything very literally. You're always on edge. Relax."

"Well, I do like reaching conclusions for myself, occasionally," I said. "We were saying that domestication basically consists of controlling reproduction — yes?"

"Yes," he admitted, digging around with his chopsticks in the remnants of a completely dismantled rice ball.

"And who does control reproduction?" I asked.

"Who do you think?"

"The market — hence the market is our owner," I conclude.

"No," said the palaeontologist, pointing his chopsticks at me.

"Why can't young people have kids?" I persisted. "Because of their low wages, job insecurity, the cost of housing ..."

"It's not quite so clear-cut to me."

"Well, that's what I see."

"The Swedish don't have the same problems, and they aren't having kids either."

"Capitalism is generally bad for having kids."

"I suspect it's a little more complicated than that," said Arsuaga in a low, thoughtful voice. "I think what you're saying is true, but it isn't the whole truth. It seems to me that if people could have all the children they wanted, the average number of children in a Spanish family, rather than being 1.2, would maybe just go up to something like 1.6."

"Well, at least you admit that the lack of work, low wages, housing, et cetera, are a part of the truth. What else is there?"

"This isn't something ... I'd need some statistics in order to give an opinion. And to give it some proper thought. I don't like giving an opinion without knowing what I'm talking about. Ortega said that birthrates reflect the whole mood of a society."

"Of course, pessimism about a lack of prospects doesn't encourage people to have children, and right now, young people are very pessimistic about their future."

"When I had my first son, the one who's about to get married, I could only get a job earning the equivalent of €800."

"But there was one thing inscribed in our minds: that we would be better off than our parents."

"Possibly, but rich people aren't having kids either. I doubt very much that even if you gave all the incentives in the world for people to have more children, that the average would go as high as two, which is the replacement fertility level. There simply must be more variables."

After having a ginger ice cream and a coffee, we went out onto the street, which was extremely lively.

"You still haven't told me," I reproached him as we headed for the Puerta del Sol, "who our owner is."

"We haven't got an owner, because nobody's ever domesticated us. We've domesticated ourselves. The danger is of somebody taking advantage of this docility of ours, because if we've become like little children, a tyrant could come along and ..."

"Oh, so we've turned ourselves into children," I remarked, ironically. "Children who, through their knowledge, have discovered penicillin and invented airplanes, and travelled to the moon and created the internet …"

"Yes, but children know a huge amount," he replied. "An eleven-year-old child already has the same brain as an adult. If you aren't a great mathematician by the age of eleven, you're never going to be. Chess masters are getting younger and younger all the time."

"So domestication doesn't imply any cognitive losses?"

"An eleven-year-old child can do calculus. What we don't have at that age is social intelligence."

Once we had reached Sol, which was bustling with people, the palaeontologist stopped and said: "Look how many docile people, all together in one place."

I looked out over the scene, and had to acknowledge he was right.

"Domestication isn't something that's planned," he added. "It's a circuit. In biology, everything works on the basis of feedback loops. You shouldn't see evolution as being like an arrow, but rather like a wheel. The wheel revolves, but simultaneously moves forward. We grow increasingly docile all the time. Since we are docile, we select for reproduction those who are more docile still. And since we select for reproduction those who are more docile still, we grow more docile, and so on in succession."

Having left Sol down Calle Esparteros, we headed toward the Plaza Mayor, and soon found ourselves outside the Ministry of Foreign Affairs.

"I've brought you here," Arsuaga said, stopping outside the front of the building, "because this palace, which was built in Felipe IV's time, was then a prison in which many well-known prisoners were held, General Riego among them. He was taken from here directly to the gallows on Plaza de la Cebada, where he was hanged. Luis Candelas, the infamous bandit, was also put in here, and likewise executed on this very spot, though he hadn't killed anyone. This was the prison. What am I trying to tell you?"

"What?"

"That selection, in the human species, has been carried out via capital punishment. In other words, it isn't the females of our species who have done it, like among the bonobos, because there have never been any power-wielding female alliances. In our case, the community has rid itself of aggressive individuals by either putting them behind bars or executing them to stop them from reproducing. Dead people don't reproduce. We've spent thousands of years executing individuals who were not prosocial. The self-domestication of the human species, according to Wrangham, the primatologist I mentioned before, really needs to be seen as the work of the species as a whole. End of story."

"There's still something in the idea of self-domestication that eludes me," I said.

"What?"

"The intensity of our herd instinct constitutes a problem for people being different. Yes or no?"

"Yes."

"And yet, it's thanks to those who are different that society progresses. The person who's different, on the one hand, is a danger, but on the other he or she's absolutely crucial for any advancement. Think of Galileo, for example."

"The hardest thing in human society," Arsuaga pointed out, "is to go against the flow."

"But if there isn't someone who'll go against the flow, we'll always stay in the same place."

"Well, yes, but that someone is fucked. The dissenter will always pay a price. Galileo paid a price. There's a price for dissidence. The herd instinct is very strong in the human species, Juanjo. You can see this very clearly among children, who are still more nature than they are nurture. They all want the same brand of trainers. They're more fearful than adults about being excluded from the group. How have we reached such levels of herd instinct?"

"By selecting those with the strongest herd instinct," I said, giving in.

"Precisely."

TWELVE

Confidence in paternity

On that Thursday in November, I woke in a state of euphoria. The day itself, however, dawned wretched. Since the earliest hours, it had been raining — a rain that was dirty and incredibly light, like grey flour, which blurred the outline of people and buildings. I went out to the corner to get the paper, and returned with my clothes drenched and my morale at rock-bottom. Later, I went to the neighbourhood post office to send a registered letter, and the woman who served me had swollen eyelids as if she'd just been crying. Since I was unable to make Thursday take an antidepressant, I took a spoonful myself of cough medicine with codeine that I hoard like treasure in my bedside table. Miserable old Madrid, I said to myself, was not going to infect me with its woes. Legal opiates are here for a reason.

The palaeontologist had asked to meet me at the entrance to the La Latina Metro station, but he had missed his stop and arrived late, feeling fluey as well. The idea had been to take a stroll around Lavapiés and to have lunch at an Indian restaurant.

"Why Lavapiés?" I asked.

"Because it's a multi-ethnic neighbourhood," he said, "and I want to show you the richness of the human species."

But the only ethnicity wandering the streets was the one represented by us, two Caucasian men of a certain age, one of whom — yours truly — was carrying a collapsible umbrella that was ridiculously open, because it wasn't raining vertically downwards, as is customary, but rather the rain was enveloping us like steam, a chilly steam.

"If you spot a pharmacy," said Arsuaga, "let me know. I need to get better in time for Saturday, for the International Cross-Country race in Atapuerca."

"You're doing an International Cross-Country race in Atapuerca?"

"Of course. It's famous. Runners come from all over."

"You'll be fine by Saturday," I said, encouragingly.

"What makes you say that?"

"Man, it's only a cold."

We went on walking down an empty street, which filled the learned man with frustration and rage.

"It's usually teeming here," he complained.

Finally, as it was getting very late for our lunch, a lateness that was becoming discernible from my mood, we went into an Indian restaurant that was empty, as well as dark and cold, and told the waiter to bring us the first thing he found in the kitchen, as we were about ready to faint. While we waited for the food to arrive, the palaeontologist told me that Lourdes, his wife, had broken her fibula.

"Me with this flu, and her with a broken fibula, in a wheelchair. What do you think of that?"

"That misfortunes always come in pairs. How was your son's wedding? Were you a big hit with your speech about love?"

"Ah, yes, the wedding was lovely, thank you."

At that moment, a Japanese couple came in, a young man and woman, and sat down at the far end of the room.

"Do you know why Japanese people's eyes are shaped differently to yours and mine?" he asked, nodding in their direction.

"I've no idea," I said.

"Make an effort."

I made an effort.

"I've no idea," I insisted.

I wasn't prepared to talk till I'd got some hot food inside me.

"But you will agree," Arsuaga said, "that there are only two possibilities: one, that it might be the result of ecological adaptation; two, that it lacks any ecological value."

"Everything in life has some ecological value, doesn't it?" I said enthusiastically as I spotted the waiter approaching the table carrying a gigantic tray covered in victuals.

They'd prepared a sort of varicoloured Indian tasting-menu for us, served in little pans that seemed to have come straight off the fire. There was chicken tikka masala, small meat samosas, basmati rice, a dish I couldn't name that was like fried vegetables woven into a kind of basket, and curried prawns and ladies' fingers — all of it accompanied by that thin, crunchy bread that gets toasted on an iron griddle (chapati, I think it's called), which I decided to dip in a red sauce that was spicy, but not excessively so. I was revived at the first mouthful, and turned euphoric at the second sip of a pale and frothy beer, also Indian, in which, if you paid attention, you could taste the hops. All of it was first-rate. This felt very much like happiness.

"Such a treat, eating like this when you're hungry, isn't it?" I said to the palaeontologist.

"Great, yes," he agreed. "But don't get side-tracked: we were considering whether or not the shape of Japanese people's eyes has an ecological value; that is, if they are the result of adapting to a certain environment."

"And I'd said that everything in life has some ecological value."

"In that case, tell me what the use is of having eyes shaped like that?"

Arsuaga, who had also been revived by the curry, was giving me a mischievous look. *Gotcha*, he was saying to himself.

"Your cold's gone," I observed.

"I do seem a bit less congested," he said, surprised.

"That's because you're trying to wind me up. Winding people up makes you better. I had a spoonful of syrup with codeine this morning, and codeine makes me very sensitive. I can sense someone trying to wind me up at two thousand paces."

"It's true, I am in that kind of mood, but do stay on track. What use can eyes that shape possibly have?"

"Well, we know why it's useful for black people to be black."

"The colour of people's skin is one of the few differential features that can be explained by adaptation to the environment. Melanin protects against the sun's UV. But forget about human beings for a moment. What function does a peacock's tail serve? We've talked about this before."

"It's useful in courtship. Have half of that samosa — it's to die for."

"There you are: the tail of the peacock is not the result of an adaptation to the environment. Furthermore, from an ecological point of view, it's a disaster, because it gets in the way."

"Well, they get to screw because of it."

"And this is precisely the point I wanted to make: in the animal kingdom, we find certain features that have an adaptive, ecological value — that are to do with survival — and others that are only to do with reproduction."

"So these features can sometimes be in conflict ..."

"Sometimes."

"And how does that get resolved?"

"Biology is full of temporary arrangements, trade-offs."

"Stopgaps?"

"Not stopgaps, no. Compromises. Imperfect solutions."

"OK. So there are some features that can be explained by natural selection, and others that are explained by sexual selection."

"Now you're getting it."

"And the shape of Japanese people's eyes would be the result of sexual selection."

"What is for certain is that no one has yet managed to identify any adaptive value for them."

"Shall we ask for more curried prawns?"

"I'm full already, but order whatever you like, and stop interrupting me. I keep losing my thread."

I restrained myself, and assumed an attentive expression.

"We all come from the same place: Africa," he continued. "From there, there was a dispersal that gave rise to the Chinese, Indians, Australians, Europeans ... Do you follow?"

"I do," I said, "and it seems comparable to what happened to Indo-European, which led to languages as apparently different as Spanish, English, and Polish ... which nonetheless, in their deepest structure, are identical."

"It's comparable, yes."

"But it also reminds me of the story of the inhabitants of the tower of Babel, who all spoke the same language until God muddled their languages because they'd tried to build a tower to reach up to heaven. From then on, divided into their different linguistic groups, they scattered across the Earth, leading to different cultures."

"OK," Arsuaga conceded, nodding patiently. "So, along these different routes of dispersal I was talking about, the features that would go on to distinguish all the different peoples of the Earth were gradually selected. This is why we are the same deep down, but different at surface level."

"But do we really vary all that much?" I asked, doubtfully.

"I want to assume that you would be able to tell a gentleman from Cuenca from a Japanese man."

"Of course."

"The person who served us our food was clearly from India, right?"

"Yes."

"You and I are descended from individuals who, long, long ago, like everybody, had dark skin. What happened to those who didn't have light skin in the cultural group from which we hail? They didn't make it as far as the modern day. Why not? Because along the way it was those with light skin who were selected."

"Sometimes you must think me an idiot, but you shouldn't believe that all this is so very easy to understand," I said, dipping some bread in the sauce.

"That's precisely why, because it isn't easy to understand, I'm explaining it to you slowly. By the way, that sauce is very spicy — it won't agree with you."

"I like very spicy sauces."

"There used to be a card game called *The People of the Earth*, or something like that, which featured an Inuit family, a Jewish family, a Romani family, et cetera."

"Yes, I had that game."

"Do you remember the Inuits standing next to the igloo, dressed in those beautiful furs of theirs?"

"Yes."

"Fine — well, clothes have an adaptive side to them, in that case to protect the wearers from the cold. They're functional, let's say. But they're also to do with taste."

"Did Darwin say that?"

"I'm saying that. And, on top of that, I say that what's true about outer garments could also be applied to certain physical features of different ethnicities."

"What we used to call races, or breeds?"

"Breed is what vets say. Use 'ethnicities' instead, or 'peoples of the world'."

"OK."

"Why does the Indian man who brought us our food look different to us, and to that Japanese couple?"

"Why?"

"Because Indians liked those particular features, and, via sexual selection, they have gradually reinforced their presence."

"So the shape of a Japanese person's eyes would be an aesthetic choice?"

"Could be, given that they appear to lack any adaptive value. Why does the male grouse have the plumage that it does? Because the female grouse finds it appealing. All the peoples of the Earth consider themselves to be the good-looking ones. In order to reproduce, one has to find a partner; and in order to find a partner, one has to be good-looking."

"Or have the gift of the gab."

(*Like you*, I was about to add.)

"That's another story. Don't forget this, because it's important: the secondary sexual characteristics, which distinguish men from women, have to do with the choice of a mate, and they've been selected over the course of evolution, but they lack any adaptive value. I can't stress this enough; it's crucial that you understand it. A woman's breasts have no use in nature."

"Sure they do — they're for breastfeeding babies."

"Chimpanzees breastfeed as well, but their breasts don't draw one's attention. All mammals have breasts."

"You're talking about breasts that stick out?"

"And bottoms. Breasts and bottoms that stick out, among other things."

"Right. We saw that when we visited the Prado."

"All the secondary characteristics that go to distinguish men from women, all — without exception — have to do with the choice of a mate. They have been chosen, as it were, with reproduction in mind. And they possess an enormous power, because you'd be able to distinguish a man from a woman in every single facet of them, even down to their eyes, and even if they were wearing a veil. Tell me, could you not distinguish a man from a woman by his feet?"

"I'm not much of a foot fetishist."

"Even so, if I presented you with a man's foot and a woman's foot, you'd be able to tell them apart."

"Perhaps."

"Therefore, this power, that of sexual selection, must be extremely strong. It's a serious, serious business. Write this down, make a note of it: sexual selection. It's because of it that we get Chinese people, Indians, Japanese, Australians."

"Shall we ask for a coffee?"

"I'd rather get out of here and stretch our legs a little. Plus, I need to find a pharmacy."

"Except you're practically cured now."

"Just in case. The race on Saturday is very important. I've been in training for months."

Out on the street, where the day was still murky, the palaeontologist set aside secondary sexual characteristics to focus on the primary ones.

"The primary ones are those that relate directly to reproduction," he said. "Penis and scrotum in the case of the man, and vulva in the woman. These are *external* features."

"Right."

"Some people try to make the case that men have penises while women have vaginas, as if the vagina were equivalent to the penis. But the vagina is an internal organ; I don't know why it gets associated with the penis. The equivalent of the penis would be the clitoris, which is also an erectile, cavernous appendage; it gets bigger with sexual stimulation, because its cavities fill with blood. OK?"

"OK. Men, penis and scrotum, and women, vulva. Noted."

"Or cock and pussy, as you prefer. The primary external features."

At that moment, the palaeontologist spied, on the opposite pavement, a very brightly lit-up establishment that looked like a chemist's. When we approached, it turned out to be a sex shop.

"What a coincidence — here I am, talking about cocks and pussies ..."

"That's a Jungian synchronicity," I said. "You're talking about something, and it appears to you."

"Let's go in," he said, forgetting all about the pharmacy. "That way we can combine theory and practice."

I hesitated when I saw a girl behind the counter. I was embarrassed, but the palaeontologist gave me a shove.

"Fine," I agreed. "But we're telling the saleswoman we're anthropologists."

"What for?"

"I don't think two old men just nosing around among all this gear would look great."

The palaeontologist threw me a pitying look, and opened the glass door.

We didn't need to introduce ourselves as anthropologists because the young woman, who was clearly well-read, recognised Arsuaga at once.

"I'm explaining something to this man," he said, gesturing at me with a long-suffering look. "And we've come to look at cocks. Have you got any cocks?"

"Realistic or abstract?" the girl asked.

"Realistic. The more realistic, the better."

She led us to the back of the establishment, and took an erect penis off one of the shelves, along with its scrotum, which looked totally real. The palaeontologist held it in his hands, satisfied.

"It's very good," he said, "because it includes the testicles. Have you got any unattached scrota?"

"Not unattached scrota, no," said Raquel, which turned out to be the saleswoman's name.

"I'll just have to make do," said Arsuaga. "First, the biology," he added, looking at me. "OK?"

"OK."

"Can I stay and listen?" Raquel asked.

Having agreed, the palaeontologist held the member up, till it was at our eye level.

"There are two things here," he said, "the size of the penis and that of the testicles. We'll start with the testicles, because testicles relate directly to social biology. Some species are monogamous, others are polygamous, and others still are either promiscuous or solitary. The orangutan, for example, is solitary. Social biology is determined by the genes. It isn't as though the gorilla says 'I want to be polygamous', but, rather, that's what its biology dictates. So the size of the testicles reflects what we call 'sperm competition.' Make a note of that, Juanjo: 'sperm competition.'"

"Noted."

"Sperm competition," repeated Raquel in turn, as if to memorise it.

"Sperm competition," Arsuaga continued, "happens in species in which different individuals' sperm compete to fertilise an ovum. There is a female in the group, and she's ovulating. That means there's an ovum up for grabs, you might say, one that's available to be fertilised. And there are species in which many males compete for the female carrying that ovum."

"In our species that doesn't happen," Raquel pointed out.

"Not in our species. Of course not," Arsuaga conceded. "Let's imagine a female chimpanzee. A female chimpanzee comes on heat — technically, she is 'in oestrus' — for one month every four years."

"Every four years! You're kidding!" exclaimed Raquel.

"That's the sex life of a female chimp," Arsuaga confirmed, with a look as if to say, *What can you do?* He was still holding the very life-like penis, complete with scrotum. "So, during that month, she's capable of copulating with ten males in a single day."

"Awesome!" exclaimed Raquel. "And the rest of the time?"

"Well," said Arsuaga, "she'll be pregnant for eight months, during which time she won't ovulate, which she also won't do for the three years that she's breastfeeding. So, what with one thing and another, that's four years without sexual activity. Understand everything so far?"

"We do understand, but it's kind of sad," said the girl, with feeling.

I felt like I was becoming invisible beside the young saleswoman's limitless curiosity and the teacherly compulsion of the older professor.

"But when a female copulates with large numbers of males," the palaeontologist continued, addressing Raquel, "the sperm compete to fertilise that ovum. And only one of them will be successful. Consider the fact that, in one of our normal ejaculations, the sperm number in the hundreds of millions."

"How many hundreds?" asked Raquel.

"Three or so. Do the maths. Ten copulations a day, over the course of a month."

"And three hundred million spermatozoa with each copulation!" she concluded admiringly.

"Sperm competition," Arsuaga concluded, "is nothing if not unforgiving. The male who produces the most sperm has the greatest chance of his genes forming part of the infant-to-be. And that's what it's all about, perpetuating the genes."

"Of course," I said, but too timidly, failing to catch the palaeontologist's attention. Or the young woman's.

"Chimp sperm," continued the sage, "as well as having a head and a tail, in the so-called middle part — which is where the mitochondria are — have energy-producing organelles. This is known as 'the fuel tank,' and in chimp sperm, this tank is particularly large. But in terms of what we were saying, the size of the testicles is a good indicator of the level of sperm competition in a species."

"Size is an appealing characteristic, then, for the females," Raquel deduced.

"I don't know if it acts like a secondary characteristic," Arsuaga said doubtfully. "For now, it's primary. Gorillas, on the other hand, live in groups in which there's only one male, the silverback. Lots of females and one male. So you don't get any sperm competition there, because when a female's on heat, she's only got one male available. Do you follow?"

"I do," I said, trying to get myself noticed.

"So, how big are gorillas' testicles?" Arsuaga asked the girl, as though I didn't exist.

"Small," Raquel got in first.

"Small," I repeated like an echo.

"That is, gorillas, even though they're really huge, have these ridiculous little testicles," said Arsuaga.

"That's fascinating!" said Raquel. "I've got to go take in some merchandise now, but I'll be right back. I'm going to be hanging around you for as long as you're here. Let me know if it bothers you."

"No, no," Arsuaga and I replied in unison.

"That young lady," said Arsuaga confidentially, as she went off, "would make a tremendous student, because she has curiosity. Curiosity is everything, but it isn't easy to find people who are curious, even in a university."

"Right," I said.

"So," he said in summary, brandishing the penis in the same

way certain politicians brandish copies of the Constitution, "a chimp is smaller than a human, but it has testicles the size of hens' eggs."

"And ours?" I asked, as though I didn't know this from personal experience.

"Ours are the size of, well, nuts. And think for a moment just how big hens' eggs are."

"And an orangutan?"

"An orangutan is a very special case. They're solitary, but when a female comes on heat, they find her straight away and copulate. Sperm competition doesn't happen with them, either. And, of the cases we've considered, it's orangutans that have the smallest testicles."

"Among humans," I concluded, "there's zero sperm competition, of course."

"There was in the long distant past. There isn't now, because we form stable couples. There's an expression you'll like: 'paternity confidence'. Note that down, too."

I noted it down.

"Among chimps," Arsuaga continued, "there's no way of knowing who the father is, meaning that this confidence is very low or non-existent. It could be anyone. Whereas with gorillas, it's very high. What would you say the level of confidence is among humans?"

"Low, as you see from those fathers going to great pains to have their children take their surnames. As they say, show me what you boast about, and I'll tell you what you lack."

"But how confident are you that your children are yours?"

"A hundred per cent."

"And what about in Spain generally?"

"I don't know. Right now, off the top of my head, and

from a few things I've read, I'd estimate that 20 or 30 per cent mustn't be the children of their official parents."

"No, no. It's much lower than that. Less than 10 per cent. The figures we see are about 2 per cent. The level of confidence among humans is very high, but not only here: among Kalahari bushmen, too. In any human society, you see children with a couple, for example, and it's a safe bet that he's the father. This is one of the keys of human sociability."

"Right."

"But let's go back to anatomy," he said, testing out the flexibility of the penis. "We don't have the penile bone that is found in lots of animal species, and particularly in carnivores, for example."

"That thing about the penile bone does just freak me out a bit," I said, "I keep thinking it might break."

"Chimps have it, but it's very small, almost vestigial."

"But we don't," I insisted, to reassure myself.

"We don't, because in certain evolutionary branches it's been lost."

"And the bone is flexible?"

"No, it's rigid. It is the same length as the penis when flaccid."

"And when the penis is erect, then it occupies, what, about 10 per cent of the total?"

"Maybe not quite so much," said Arsuaga.

"And it absolutely never breaks?"

"No, it never breaks. The penile bone of a bear is quite considerable. What I was going to say," he said, while showing me the life-like latex member again, "is that the length of our penis is about the same as a chimp's. In girth, however, we're the winner among all the primates."

"Why?"

"Well, that we don't know."

"Wouldn't it have something to do with the width of the vagina?" asked Raquel, who'd just come back.

"Possibly," conceded Arsuaga. "Some people claim it's to do with stimulating the clitoris, but we don't have a definite explanation. In any case, write this down as well, Juanjo: the human penis is much wider than that of any other primate. Much wider. Some say it's that shape in order to dislodge the sperm from the previous copulation."

"It would act like a suction pump?" — this girl was lightning fast.

"Precisely. But a theory like that contradicts the size of the testicles, because the size of our testicles indicates there's no sperm competition. And if there's no sperm competition, nor is there any point in dislodging the results of the previous ejaculation."

"Right," I managed to get in first.

"It may be that it has more to do with the diameter of the vagina," continued Arsuaga, "because the head of a human child is larger than that of an infant chimp. Anything you're unsure about?"

"No," I said, which settled the matter.

Arsuaga put the penis back on the shelf it had come from, and looked around.

"Out of all this stuff," he said, "the only thing that makes any sense to me is the lingerie. You?"

"Same here," I replied.

"We do have artificial vaginas, too," said Raquel, as if she was afraid we might be about to beat a retreat.

"Curiously enough," said Arsuaga, "that's what we were

talking about, penises and vulvas, when we saw the shop."

"A Jungian synchronicity," I explained to Raquel, "which is a ..."

"I know what a Jungian synchronicity is," she replied, a bit put out.

"We thought it was a pharmacy," continued Arsuaga, "because of the shop window and the lighting. I need a flu remedy because I'm running in the Atapuerca cross-country race on Saturday, and I'm coming down with something."

"Everything to do with sex used to be kind of kept in the dark, right?" said the young woman.

"Certainly," I said, aware of having emerged from those dark places myself.

"But not anymore!" she said. "Now everything related to sex is brightly lit, joyful. So, do you want to see the vaginas?"

Arsuaga gave me a questioning look.

"Alright," I agreed, so as not to look like a prude. And she led us to the part of the store where those bits of body topography were displayed. It turns out that they were, mostly, exact replicas of the vaginas of famous porn stars. Since we didn't know the porn actresses whose names Raquel enumerated, they looked to us like normal vaginas, assuming there's such a thing as a normal vagina.

"And do these actresses charge for copyright?" I asked, holding one of the vulvas, which felt like human skin.

"Of course!" said Raquel, as if to say, 'What did you expect?'.

I suddenly found myself in a room strewn with dismembered body parts. Everything that was appealing about the store, with its lighting, its décor, its joyfulness, its background music, its polymers plastics, was all crashing down on top of me. I wanted to go, to flee, but Raquel and the palaeontologist had become engaged in in a curious argument. According to her, when a number of women live together, their periods become synchronised, as if they were mysteriously connected.

"So people say," said Arsuaga, "but it's just an urban myth; it isn't real."

"Well, I've experienced it with my mother and my sisters. And also with my flatmates."

"It's far from confirmed," Arsuaga insisted. "They've done studies in women's prisons that contradict the idea."

Raquel seemed annoyed. I tugged gently on the palaeontologist's arm to see if we might leave, because I was starting to feel claustrophobic. The customers who came and went were giving us funny looks. When we had almost reached the door, Arsuaga stopped and addressed the young woman: "Out of interest, Raquel: in your personal experience, and that of your friends, at what point in your cycle is your libido highest?"

"My libido increases before I start menstruating, and three or four days after my period has gone."

"That's very mysterious," said the palaeontologist, looking perplexed. "Biologically speaking, very mysterious, because the logical thing would be for sexual desire to coincide with ovulation. Or not?"

"Of course," I said.

"What I notice in the middle of my cycle," added Raquel with a dreamy expression, "is that I'm, I don't know, it's like I'm more sensitive to beauty, I'm more perceptive."

At that moment, a customer who'd been nosing around called her over. I told Arsuaga I was going to struggle to connect the conversation in the sex shop with what we'd talked about over lunch.

"What's so difficult about it? We've been talking about biology, and all of this," he said, waving his hand to indicate the contents of the shop, "is biology."

"It's all cultural," I ventured.

"It couldn't be more biological."

"It couldn't be more cultural."

"A vagina, a penis, they're cultural?"

"If they're artificial, yes."

"A penis is a penis, and a vagina is a vagina," he said, as if that settled it.

"Whatever you say, but I'm off now, I've got to go to the supermarket."

"Aren't you interested in getting Raquel to explain what some of these little toys are for?"

"We'll come back another day, if you want. I've got to do my shopping."

We said goodbye to the young woman, at last, who encouraged us to come back any time we liked, and the moment we stepped out onto the street, Arsuaga sneezed.

"It's a mystery, the way this cold keeps going away and then coming on again."

"It comes and goes because it's psychological," I gave my diagnosis.

"You set great store by psychology."

"And you set great store by biology. Look, over there, a pharmacy."

Fortunately, on this occasion, what looked like a pharmacy was in fact a pharmacy.

"I'll wait for you outside," I said.

After a while, since it was taking him some time to come out, I went inside to see what was going on. The pharmacist, very patiently, as if repeating the same thing for the third or fourth time, was saying to him: "Nothing has been invented that cures a cold. I can give you something that will relieve your symptoms."

"That's fine," conceded Arsuaga, "give me something for the symptoms, because I'm supposed to be running in Burgos on Saturday."

We left the store with a box of Frenadol.

"I would have given you a better prescription than that," I said to him. "Frenadol has got pretty old now."

"I might not even take it. Are you getting the Metro at La Latina?"

"Yes."

"I'm not, but I'll walk you there. Open your umbrella."

I opened my umbrella, even though the prevailing wetness made it feel like it was raining upwards.

And that was that.

THIRTEEN

His distant footprints

The palaeontologist proposed that we take a trip together.

"We've already taken trips together," I resisted.

"We've been out on the odd excursion, but we've never stayed away from home together," he counterattacked. "You really get to know people on trips."

"I'm not sure I want you to get to know me," I objected. "Or if I want to get to know you. We could ruin everything."

"We ought to take a chance," he said.

And that was how, on 13 November, a Wednesday, I came to pack my suitcase and was standing at the door to his house at 9.00 a.m. The weather, just as on our previous meeting, was cold and disagreeable. It was raining quite unenthusiastically, intermittently, the way a child, worn out with crying, still won't give up on prolonging his protest with intermittent wails.

I informed him via the intercom that I'd arrived, and he came right down. When he saw my outfit, he burst out laughing.

"You look like you're going to a book launch at the Palace Hotel!" he exclaimed.

"I still don't know where we're going," I replied. "You haven't told me."

He was kitted out kind of like Indiana Jones. He was always kitted out kind of like Indiana Jones.

"It doesn't matter where we're going," he said. "Are you familiar with Decathlon?"

"I know what it is, but I've never been."

"I'll take you one day, so you can give your wardrobe a bit of an update. And what's with the suitcase?"

"What about the suitcase?" I asked, a bit annoyed now.

"It's pretty damned big, is what. We're only going for a night. What have you brought?"

"Things," I defended myself. "Just in case."

"Well, I've got everything in this rucksack."

As we talked, we were walking toward his Nissan, which was parked nearby. Once inside, before he started it up, I gave the dashboard a couple of pats by way of greeting.

"How many kilometres has it done?" I asked.

"One hundred and twenty thousand," he said. "And it runs like new."

"What's the model called?"

"Juke," he said. "It's got a very muscular front section. When you look at it head-on, it reminds you of a samurai's face. It's the car that saved Nissan from going to the wall."

"Right."

"And the Nissan Patrol," he added, "ended Land Rover's supremacy."

"I had no idea," I said.

We started it up and set off toward the Burgos highway. At one point, I spotted the four towers at the old Real Madrid training ground, whose upper floors were enveloped in a mist that looked like Arsuaga's grey tousled mane.

"Buildings with manes," I said, receiving no reply. On the

radio, they announced that a lorry had overturned at Terminal 4. The palaeontologist switched it off and asked me if I suffered from claustrophobia.

"It depends," I hesitated.

"The place where we're going, you can't have it."

"And where is it we're going?"

"You'll see."

When we joined the motorway, I felt a bit of unease because of the November landscape, so grey and stiff. The mist clung to the ground like a shroud to its corpse.

"What are you thinking?" Arsuaga asked me.

"I've always wondered, ever since I was little, why anything exists."

"Do you mean why there's something instead of nothing?" His gaze swept across the landscape.

"Yes."

"Well," he said, his tone turning didactic, "there was a time when everything was nothing. But nothing is very unstable and, in one of its moments of indecision, it gave rise to everything."

A great answer, I thought, that reminded me of the last lines of the José Hierro sonnet entitled 'Life':

> Nothing remains of what was nothing.
> (It was an illusion, believing all
> what was, in fact, nothing.)
>
> Surely no matter that nothing was nothing,
> if nothing more will be, after all,
> after so much all for nothing.

I imagined, besides, that the instability of nothing was emotional in nature, like my own, and that helped to reconcile me with the landscape a little.

Hi, landscape, I said silently.

"The landscape," added the palaeontologist as if he could read my thoughts, "is a document. You need to know how to read it in the same way you need to know how to read the human body."

There were a few moments of quiet, livened up by the swish-swish of the windscreen wipers moving back and forth. Then he asked me if I was a pain in the arse. I didn't know what I ought to reply, and so I just did that swaying movement with my hand that means *comme-ci, comme-ça*.

"What about you?", I asked, in turn.

"The Basque country has produced more than its fair share of pains in the arse," he said proudly, "and I am a member of this noble tradition of troublemakers, that of the Basque pain in the arse."

"Right. And there are degrees of pain-in-the-arse-ness?"

"Of course. The pain in the arse par excellence is the one who would have all the opposing sides shot because he or she doesn't feel quite right joining any of their ranks. The accepted idea is that Galileo was condemned for claiming that the Earth went around the sun, but I think people in general didn't punish him for his ideas, but for being a pain in the arse. Servet wasn't burned at the stake for arguing over pulmonary circulation, but because he was a major pain in the arse."

"Would you have survived the Inquisition?"

"I don't think so, because I've always made a pain in the arse of myself, questioning everything ever since I was young, first at home, then in school, then at university ... Darwin, who was no kind of pain in the arse at all, he got along fine."

"Turn the wipers off — it's stopped raining."

"It isn't raining a lot, but it is still raining. Does the sound annoy you?"

"Not too much. Forget it."

"Galileo," he continued, "got on very badly with the Jesuits, even though he was the most religious person around. He said, get this, that the universe was the letter God had written to man. From that to claiming that science is theology, it's no leap at all. But he just rubbed people up the wrong way. Copernicus, who had said the same things before him, died peacefully in his bed."

All of a sudden, a gap opened up between the clouds, and a tremendous torrent of light came through it.

"You just mention God, and look what happens!" I exclaimed.

"The whole thing of belonging to a tribe that will protect you," said the palaeontologist, unmoved by the miracle, "is all well and good, but it's also a hassle."

"But we all need a group to belong to, don't we?"

"Not all of us."

At that moment the gap closed up again, and the landscape went back to being gloomy.

"Just look at those autumn colours," he said, gesturing at the vegetation. "The rusty browns, the yellows, the purples, the ragged clouds … It's important not to miss autumn — good thing I thought of this trip. Look at the meadows …"

I did look at the meadows without managing to lapse into Arsuaga's mystical rapture.

"I don't know if I might be coming down with something," I said. "My ear's bothering me a bit."

"It's because we're going up in altitude. Swallow."

I swallowed.

"Can you hear me better now?" he asked.

"Yes, yes."

"The human ear," he said, picking up on the point, "is amazing for doing what it does, but it also has a lot of problems. The hammer and anvil bones used to be parts of the jaw in reptiles, and only later became parts of the hearing apparatus. Mammals like you and me are based on the blueprint of reptiles. Not that it's perfect, but for something that's been cobbled together, it isn't bad. We are made from the second-hand clothes our older siblings threw out. The placenta, for example, is based on an egg. Placentas are great, but you can't expect the same perfection of them as you could if they'd been made *ex novo*."

"Deep down," I said, "we are still fish."

"Well, yes. In fact, our lungs were the flotation devices. Our organism has been built in the same way you might put together a book: correcting this, crossing out that … We aren't the result of any kind of planned process, a design. Nature, as Darwin demonstrated, lacks purpose. Nonetheless, it is capable of creating biological structures intentionally. Nature doesn't seek, but it does find."

A disoriented bird crashed into the glass of the windscreen, giving us a fright. The bars of the wipers removed the remains of blood and feathers mixed with the rainwater.

"I think we've killed it," I said.

"Yes, poor thing," said Arsuaga. "It was a blackbird."

After a few moments without speaking, as a sign of mourning, I asked him if there is any creature in nature whose perception of death is even remotely like ours.

"Elephants and chimpanzees find death perplexing," he said. "They don't know what to make of it. In evolution, there are some branches in which social complexity appears, and others in which it doesn't. A revolution would never be possible in an ant nest, for example."

"And what about in a chimpanzee colony?"

"In a chimpanzee colony you get politics, you get alliances, you get a struggle for power, all of which would be unthinkable in an ant nest. There's no *mood* in an ant nest; they're nothing but little machines. A chimp, a dolphin or an elephant, on the other hand, are all sentient beings. They feel hunger and thirst, for example, and emotions as well."

"Ants eat, therefore they experience hunger."

"They don't have hunger; they have a thermostat. The battery in my mobile phone isn't *hungry* for electricity, but when it's running low it still tells me to plug it in."

"So elephants, dolphins, and chimps all have selves, then?"

"They don't have a self exactly, but they do have some kind of subjectivity conferred on them by their experience of hunger, thirst, or pain. You get none of this in an arthropod. Whatever it is that arthropods feel, it bears no resemblance to the experience of vertebrates."

"When you cut a living lobster in half and put it on the hot griddle, doesn't it feel pain?" I asked.

"Have you done that to a lobster?"

"Yes, but with a certain amount of guilt, even after my fishmonger told me that they lacked a central nervous system, meaning they didn't suffer."

"Damn him, that fishmonger of yours!" said Arsuaga. "He was right, though. Invertebrates don't have brains."

"What do they have?"

"Ganglions. Rest easy, I don't think your lobsters would have suffered on the griddle."

"But even cut in half, they still keep waving their legs. It freaks me out a bit, if I'm honest."

"A mechanical reaction. Pure thermostat."

"So a lobster can eat without experiencing hunger?"

"Look, an amoeba reacts to chemical stimulation. A microbe has information about the world around it, and it reacts on the basis of this information in the same way that a robot built to cut your grass would. Have you not seen it? When they're about to run out of battery, they go back to their charge point and plug themselves in. Are they hungry? No, they are in possession of information. Would we say that a bacterium has subjective experiences? We wouldn't. Why should lobsters or crabs, then? A chimp, on the other hand ... A chimp is confronted with death, and is completely disconcerted."

"You're reassuring me."

"In summary," he concluded, "bacteria do feed themselves, of course, but they don't have the experience of hunger."

"And what's consciousness for, other than for experiencing hunger and death?"

"What's it for in the case of a cow?"

"I don't know."

"I'll tell you: nothing. It's useful to a social animal like us, for politics. And don't confuse herd instinct with sociability. But hey, how great to be going on a trip mid-week, to get away from teaching, even with this rain!"

"Where are we going?"

"You'll see. Enjoy the surprise."

Arsuaga drew my attention to some copper-coloured beech groves to our right. Next, an oak wood appeared. The sun and the clouds were alternating in a sort of duel, as if somebody, up there, were having fun turning the lights on and off. I started to feel sleepy, because the purring of the Nissan Juke's engine was lulling me and also because I'd had a bad night thinking about this trip to nowhere.

"Sorry," I said, "I'm going to take a quick nap."

"No problem at all, I'm the one driving."

After some unspecified period of time, the palaeontologist woke me up.

"You mustn't miss this," he said, with the satisfied expression of someone giving a present.

"This" was the Cares Trail. I recognised it because I'd been through here many years before, when I was young, on an end-of-year trip I took to the Picos de Europa mountains with my university friends. The waters of the Cares flowed down through a tight little passageway, a ravine, like a throat that connects León with Asturias the way an oesophagus connects the pharynx with the stomach. The road, which is very narrow, runs parallel to the river, which was on our right. When I looked up to one side and then the other, I saw only very tall, irregular rocks, with vegetation that was sparse but dense in those places where it could grow. From out of the chinks of these high walls there arose, at intervals, springs of water, sometimes proper waterfalls, whose source, I presumed, was the rain. I estimated we must have been on the road for three or four hours, if not more, which had gone by in a heartbeat for me.

"You're a really good driver," I said, clearing my head.

"Thanks, old man."

The palaeontologist was indeed a calm, restrained driver. He didn't make any abrupt movements; he didn't let the car jolt; he didn't brake or accelerate in sudden bursts; he didn't mistreat the innards of the Juke, the model that saved Nissan from ruin. The gorge's hairpin bends meant he had to be constantly handling the steering wheel. We were tracing s-shapes, sketching curves, drawing sinuous lines in the deepest part of that depression in which the journey felt more mental than physical.

In this way, and without ever leaving the gash carved into nature by the erosion from the Cares, through which we were sliding, bemused, like a baby through the birth canal, we arrived mentally, but also bodily, at Las Arenas, a village in the Asturian municipality of Cabrales, right at the foot of the Picos de Europa. The palaeontologist stopped the car at a strategic location and invited me to get out so that we might contemplate the Naranjo de Bulnes, whose peak rises up like a totem pole amid the huge Urrieles Massif, at some two-and-a-half thousand metres above sea level (we were at about a hundred and forty).

The palaeontologist was levitating at the vista.

"Don't ever forget this moment," he said.

"I won't," I assured him, gawking.

"You might say that God leaves a lot to be desired as a watch-maker, as an architect, as an engineer, even as a biologist; look at all the damned beetles, after all. But as a landscape architect, he really can't be beaten. You can't disagree with me there."

I did not disagree.

"And now," he said, "let's go and eat. It's high time."

We went to a restaurant in Las Arenas, where we were being awaited by Pedro Saura and Raquel Asiaín, two of the palaeontologist's friends, who I soon learned were working on a prehistoric cave outside the town. Pedro and Raquel were dressed in overalls, presumably from Decathlon, which were mud-stained.

"We've just come out of the cave," they apologised, looking rather apprehensively at my get-up.

I could tell what they were thinking: *This one looks like he's going to a book launch at the Palace Hotel.*

To which the palaeontologist added, out loud: "We've already agreed I'm going to take him to Decathlon one day."

As a first course, we were served a giblet soup. I say giblets because it contained some shredded matter that reminded me of the stuff that used to float in the soup that, when I was a boy, we got served at home on Christmas Day, and the sight of which made me nauseous because my father used to say it looked like "the primordial soup".

"The one where life first appeared," he'd add, looking at each of us in turn, somewhere between grief and astonishment.

It was from out of there that my siblings and I, and humanity as a whole, had emerged: from a soup that was cloudy and dark, with little bits of chicken entrails and ground almond. From a pond.

Pedro Saura, an emeritus professor in Fine Arts, was more or less my age and was responsible, along with his late wife, Matilde Múzquiz, for the recreation of the Altamira ceiling, the Neocueva, located next to the original site. Raquel Asiaín, who I estimated to be about thirty, was carrying out, under Saura's supervision, a doctoral thesis on the intelligent exploitation of the wall surface on the part of Palaeolithic artists. She was researching the way they made use of the reliefs in the rock to emphasise certain parts of the figures they had drawn (the forequarters and the hump of a bison, for example).

A complicity was immediately established between them from which I felt excluded, which allowed me the appropriate distance to appreciate the pleasure that this meeting, at the feet of the Picos de Europa, on a regular Wednesday in November, prompted in three people with shared interests and knowledge. Pedro had a laugh that was open and resonant and that relaxed the atmosphere. The palaeontologist immediately went into ironic, bit-of-a-pain-in-the-arse mode, which in him is a sign of happiness. Raquel Asiaín, the most discreet of the three, perhaps owing to her being in a minority in terms of gender as well as of age, found herself, it seemed to me, halfway between being excluded and being complicit.

During the main course, which was similarly calorie-rich (fried eggs with rashers of bacon), I had a sudden attack of relevance. I call it that because under its influence, reality is turned into a Flemish painting. This involves people and objects taking on amazing hyper-real characteristics. A glass of wine, for example is transformed into THE GLASS OF WINE. A fork becomes THE FORK, and a spoon, THE SPOON. During these raptures, I feel myself wrenched out of the world of things to enter the platonic world of ideas. I fall, in short, into a mental state from which I am able simultaneously to appreciate each body not only considered in isolation, but also embedded within the whole. I see it all, including the harmonies and the bones that flow between the components of reality, whether they are animate or inanimate.

"You look preoccupied," said Arsuaga, just as I was dipping a strip of bacon into the egg.

"It's just that this piece of bacon isn't a piece of bacon," I replied, "it's THE PIECE OF BACON."

Pedro Saura gave one of what I thought of as his syntactical laughs, as they served to unite loose parts of the conversation, and explained to me that La Covaciella, which was the name of the cave where they worked and which we were going to visit after lunch, was from the same period as the one at Altamira.

"It has four magnificent bison."

"Now you know what we've come for," Arsuaga said to me. "To see bison from fourteen thousand years ago."

I gave a nod from within my attack of relevance, and turned my attention to Saura, who at that moment was saying: "My theory, which I have no way of proving, is that the creator of the Altamira bison went into the cave with a concrete purpose in mind: he was going to draw bison, which was also to be of a specific size. He got some flint blades, which he would use as chisels, and first engraved the animal's outlines into the rock: horns, beard, fur, everything ... The engraving is sometimes about a centimetre thick. Then he took a particular kind of charcoal, which you definitely don't get in the Altamira area ..."

"How do you know you don't get it there?" I asked.

"Because no traces of this charcoal appear in the pollen analyses of the site."

"Carbon from pine trees?" asked Arsuaga.

"Yes," said Saura. "Pine that grew on the Picos de Europa, which means this man knew how to obtain charcoal that didn't break down in a reducing flame. There are lines more than a metre long done in a single stroke."

"You need a lot of skill for that," Raquel pointed out.

While I was mopping up what was left of the egg and the fat from the bacon with a piece of hyperreal bread, I was travelling mentally, with no effort at all, to the era they were talking about. It wasn't that I saw the man going into the cave with a concrete purpose; it was that the man, in a way, was me.

"What else?" I asked.

"The Altamira bison, unlike the ones you'll see this afternoon, which are black, have parts in red. And that is due to a piece of complete chance. It turns out they painted over some underlying horses, from about four-and-a-half thousand years earlier, drawn in iron oxide. The artist liked the effect, and filled out the red."

"And why are you so sure it's all by the same artist?"

"Because all the bison are handled with the same method. Since the rock has a texture to it, the charcoal is left on only part of that texture, which gives us objective data about the direction of the stroke. Charcoal doesn't get left on the same pores if you do it like this, compared with doing it like that," he said, moving his hand with an imaginary piece of charcoal one way and then the other.

"Of course!" I exclaimed enthusiastically, because I was starting to see inside my head everything that was being revealed to me, as if my cranial cavity were a cave from about fourteen thousand years ago whose walls were covered in bison.

"The direction of the strokes," Saura went on, "is always the same, and runs parallel to the animals' fur, as if he were caressing them. There's not a single stroke that goes against the direction of the coat."

"That's a good indication of its being one artist," Raquel explained. "It's a marker of style."

"And why are you always talking about it being a 'he'?" I asked. "Couldn't the artist be a woman?"

A silence fell — an uncomfortable one, it seemed to me, which Saura resolved with a laugh followed by this explanation: "The Altamira painter was between a metre seventy and a metre eighty tall, a very great height for a woman at the time. He painted on his knees, and in some cases lying down, because the roof was closer to the ground than it is now. There are strokes a metre twenty long painted all in one go. It was a tall person with long arms. It was a man."

For dessert, they brought us rice pudding with a crust of burned sugar on top that needed to be broken with the tip of the teaspoon, like a pane of glass, to reach the rice beneath. While we tackled it, Saura said that Asturias had seventy painted caves in it.

"Not counting the underwater ones."

"Exactly," said Saura, "not counting the ones that are underwater, because the sea level was ninety metres lower than it is now. They're flooded."

The fact that they're under the sea, I thought, didn't mean they were flooded; maybe not. They might have formed pockets that were resistant to the water. I thought the idea so thrilling I wanted to pause on it.

"As well as the ones that must still be sealed," continued the palaeontologist. "This area has a uniquely high density of them."

"The one we're going to see this afternoon, the one at La Covaciella, is it much like all the others?" I asked.

"No, they're all unique," Saura intervened. "This one was discovered in 1994. They laid down an explosive charge to widen the highway, and opened up a hole, which is the one we're going to drop in through. No intervention of any kind has been carried out to this cave. It's exactly as it was fourteen thousand years ago."

Having drunk up our coffee and left the restaurant, we quit Las Arenas along the AS-114, and soon arrived at a kind of site hut right on the roadside, beside which we left the cars. Before going in, they provided me with protective gloves and a miner's lamp, like the eye of a cyclops, which only lit up the spot my eyes were pointed at, so my peripheral vision, once we were inside the cavity, would be reduced to zero. There was nothing inside the hut, apart from a hatch that allowed access to the cave, into which you descended vertically, as if into the deepest depths of oneself, down an iron staircase that the prevailing humidity had made very slippery. Pedro Saura and Arsuaga went down ahead of us. I followed Raquel Asiaín, whom I leaned on like a blind man onto his guide. Up above was the guard who'd accompanied us and who shut the hatch once we had passed through.

Suddenly trapped within that bubble of darkness, we moved our heads from side to side, causing a crossing of light sabres in the middle of an inconceivable darkness and an unheard-of silence, if silence could be heard.

"Now what?" I asked, just to check whether I could still hear my voice in those depths.

"Now try to put your feet where I've put mine," said Raquel. "Lean on your hands when you need to, because the ground is really slippery."

I quickly learned that leaning on your hands was a euphemism for walking practically on all fours, since the terrain, which was very irregular, was full of recesses and protrusions that prevented normal locomotion. When I looked up to try to make out the depths into which we were headed, I saw uneven patches of an organic nature, like the ones you find in the gullet of an ogre. It felt like a cave made of flesh, carpeted by remains of tonsils and mud, a lot of mud, which clung to my smart shoes and my book-launch-at-the-Palace Hotel outfit.

The oesophagus down which we were sliding like ridiculous boluses down a giant's throat opened into a kind of hall that could only be reached by scaling an almost vertical ramp, one-and-a-half or two metres high. Fortunately, somebody had hung a knotted rope that facilitated the operation, which consisted of holding onto one of those knots, resting your foot against the wall and pushing yourself upwards hard until you had conquered the next knot.

When the goal had been reached and a vertical position resumed, I aimed my cyclops stare at the wall on the left, and was met by a hyperreal bison that had been waiting for me for fourteen thousand years, a whole life. The torch beam first took in the image in its entirety, which I thought immensely powerful, and then moved around its outlines, in which incidental features stood out, like that of the face (extremely human), the horns (so very delicate), the beard, the mane, the legs (really elegant), the eye (very dynamic) … The shading that was to emphasise the trunk, meanwhile, was as effective as it was severe. The mixture of complexity and simplicity was admirable. Both qualities were fiendishly intertwined, as in an alloy in which it's impossible to isolate the component parts. I was shaken to the core by the idea that I was a hundred and forty centuries away from this artist, and yet right beside him in terms of physical distance, because I was in the same space in which he'd worked, perhaps treading in his distant footprints at this very moment.

Just then, Pedro Saura turned on the floodlight they used to take photographs, revealing the whole group, which comprised four bison: three of them looking toward the left, and the other, separated from the trio by a crack, to the right. The paintwork seemed fresh, as if it really had been done yesterday. I asked how it was possible for that vitality to have been preserved.

"The humidity carbonates the lines and affixes them to the walls," said Raquel.

After a few minutes of ecstatic contemplation, Saura spoke: "One of the conclusions I've reached after having visited more than a hundred caves is that whoever they were, the people who created these marvels, they were unique personalities within their communities. They did some really amazing things kind of out of nowhere. Look —," he added admiringly, pointing with his index finger at the profile of one of the bison — "you see the drawing has been done over an earlier engraving, and the engraving cannot be corrected. If you get it wrong, the groove remains. As if that wasn't enough, the figure is framed by other parallel grooves — you see? — that look like they're made with a kind of comb."

"This wasn't done by any old person," said Arsuaga.

"They clearly had real knowhow," confirmed Saura. "They were professionals."

"Look," added Raquel, "how the artist has taken advantage of a part of the rock that's bulging out to give the chest some volume."

"Is it possible," I asked, "that the shape of the rock prescribed the background — that is, it determined which animal was to be drawn?"

"I don't think so," said Saura. "I think they had a very clear idea of what they were going to draw. It's another matter if they made the most of the wall surface to draw attention to certain parts of the body."

"I wonder," said Arsuaga, cutting in, "if the act was all just about producing the bison."

"What do you mean?" I asked.

"As in, maybe this wasn't decorative art, with the focus on

making something that would last, but a ceremony, all about the actual moment it took place. That would mean they weren't bothered about painting one animal over the top of another, like they did in Altamira. Once it was painted, there was no point to it, meaning another could be done on top."

"We'll never know," declared Saura.

There we were, in short, four individuals from the twenty-first century on the same piece of ground where some of our ancestors had trodden fourteen thousand years ago, strangely linked to them through this image that was straining to escape its limestone canvas.

"Just imagine the movement these figures would take on," Saura continued, "by the flickering light of an animal-fat lamp, which is what they used for illumination."

As to the reason for the human appearance of the bison faces, the answer was that we haven't the faintest idea.

"We don't know," Saura concluded, "whether these pictures were manifestations of art for art's sake, if it was a propitiatory activity related to hunting, or if they were associated with fertility. It could be a bit of everything, but personally I think it's just superb that we have no way of ascertaining it, that it remains mysterious."

When we left the cave, it was already totally dark outside. As we headed back to Las Arenas to check in to the hotel where we were to spend the night, the wavy outlines of the landscape formed clumps of opacity that opened out into the prevailing darkness beyond. Here, I thought, is some prehistoric wind and weather.

"These," said Arsuaga, in reference to those artists from the past, "were the cream of the crop: they spent all day going from place to place, they had a very varied diet, and larger brains than current-day humans. They were just as clever as us, if not more so. And the best part: they were incredibly vain. A group of humans has never been so full of themselves. They spent the whole day painting their bodies, decorating themselves to look good: pendants, bracelets, and necklaces using animal claws and feathers, tattoos ... To my mind, this represents a collective mood; if people are depressed, they let themselves go. Skeletons have been found in Russia with incredible numbers of ivory beads sewn into their garments. The fabric of the garments hasn't survived, but the beads have, and you wouldn't believe how much time it takes — hours, months, years — to make such adornments. They devoted a lot of time to personal cleanliness. They saw themselves as good-looking, they felt good-looking, they knew they were good-looking. And just see what they came up with when they put their minds to drawing."

Later, in bed, as I rushed, eyes closed, into myself, I was really falling back into La Covaciella and once again seeing the paintings in a hallucination that has not yet stopped, because the cave is within me still.

That night, it snowed.

The following day, after having breakfast and wrapping ourselves up warm, we went to say goodbye to the Naranjo de Bulnes, whose peak looked as though it was covered in a lid of recently burnished silver like that of a classic fountain pen. I saw, in my fantasy, a colossus taking off the lid and using the pen to go over the outlines of that immeasurable massif.

Two days later, I received a message from the palaeontologist. It said: "Our relationship has survived the trip, but I can't say whether I know you any better."

I replied that he remained a mystery to me, too.

As for my book-launch-at-the-Palace outfit, it was rendered unusable, but we still haven't gone to Decathlon to replace it.

FOURTEEN

Not as simple as it looks

One day, the palaeontologist took me to the Palomeras School and Cultural Centre in Vallecas, where his friend Mario García worked.

"You've got friends everywhere," I said.

"Does that seem like a bad thing to you?"

"Did I sound like I was reproaching you?"

"Slightly, yes."

It was January; it was still cold.

I was a little depressed, not for any reason but because it's in my nature. Depressive people hate vital people, out of envy, and the palaeontologist is one of those who is always doing well. You might find him pissed off, but never sad. *Perhaps*, I thought, *he fights sadness with annoyance.*

"Do you never get discouraged?" I asked him once.

"No way," he said. "I'm a follower of Unamuno. I have a tragic sense of existence."

"Well, you don't look very desperate."

He was at the wheel of his Nissan Juke, and he turned toward me as if to say, *Well, what can I tell you?* Then, for a moment, he seemed like a man filled with panic. Filled with panic at not being good enough at what he does, whatever it was he did. In his panic I saw a reflection of my own, and sensed why we had formed that rare society of ours. On the radio, Luz Casal was singing "You play at loving me, I play at you believing that I love you ... And I don't care at all."

That day, I didn't care at all. The palaeontologist, meanwhile, was eager to show me an experiment.

"We're off to see some three-year-old children. You're going to be amazed."

"Amazed at what?"

The classroom tables, low and hexagonal, were each occupied by six children. There were about twenty children altogether. Arsuaga's friend introduced us to the teacher (Maribel), with whom we chatted by the door, unseen by the kids, who hadn't noticed our arrival.

"This," Arsuaga explained, "is about trying to establish what age we are when we acquire what in evolutionary psychology is called a 'theory of mind'."

"And what does that consist of?"

"Of you realising that other people have ideas in their heads, the same as you do, and then establishing hypotheses regarding those ideas. This is fundamental, because it's the basis of manipulation and deception. Animals can't lie, because they don't have a theory of mind. Got it?"

"I think so," I replied, trying to guess with my theory of mind what the palaeontologist was hiding in his head.

"When you have a theory of mind," he added, as if he'd read my thoughts, "you spend your life imagining what the other thinks. If what you believe they think isn't of interest to you, you try to impose a different idea on them."

"How terrible!" I said.

"Manipulation," he continued, "can be good or bad, and in general it happens completely unconsciously. But the fact you're aware that someone else believes something that's incorrect, or something that isn't to your advantage, presupposes that you have acquired a theory of mind."

The kids went on doing their thing while the palaeontologist, the teacher, and I plotted behind their backs.

"Imagine," continued Arsuaga, "that I go into this classroom with a chocolate cake, and that at one end of the room there's a box, and there's another at the other end, and they're both empty. I put the cake in the box on the right, and I go away. Soon after that, the teacher takes the cake and, with everyone watching, hides it in the box on the left. Then I come back in to get it. Where do the children think I'm going to look? In the box where I hid it, or in the one where the teacher hid it?"

"In the one where you hid it," said the teacher.

"And where do you think they'll be looking — at the one where I think it is, or in the one where it really is?"

"At the one where you think it is, so as not to give you any clues," I said.

"Well, if that happens, it means these children have already acquired a theory of mind. They know that I, like them, have a mind, and because of that it's possible to manipulate me. I am capable of being deceived. If it were otherwise, they'd expect me to look for the cake where the teacher had put it — that is, where it really is — and they would look in that direction and not realise they were giving away its true location."

Arsuaga then asked Maribel if the pupils had a favourite toy out of all the ones we could see around the place. The teacher pointed out a spaceship, almost half a metre long.

"It arrived yesterday, and they all want to play with it," she said.

Arsuaga took it, gave the teacher some instructions, and then called for the children's attention. They turned toward him, suspicious of his possession of the toy.

The palaeontologist started to roam around the classroom, moving his legs and arms in an exaggerated fashion reminiscent of those evil dolls in a puppet theatre. Then, to bemused looks from the children, he hid the spaceship in a cupboard and left the room with a mischievous expression on his face.

The palaeontologist is a real performer.

Then Maribel went over to the cupboard and, gesturing for the children to be quiet, took out the toy and hid it in the furthest corner of the room, behind a bookcase. A few seconds later, Arsuaga came back in, looking as though he was about to retrieve the spaceship. The children, confirming that at the age of three they had indeed acquired a theory of mind, looked at where the palaeontologist expected to find it, not at where Maribel had hidden it.

Stunning.

The theory was stunning, and so was the performance given by the palaeontologist, who had enjoyed the dramatisation as much as the kids, if not more so.

"Until not long ago," Arsuaga explained to Maribel and me, "people believed that a theory of mind wasn't acquired until a child was four years old — but this lot are only three, and they've tried to trick me."

After the experiment with the infants, the palaeontologist asked for us to be taken to a year four class, where the children were nine years old.

There he projected onto the whiteboard one of the hyperreal bison from fourteen thousand years ago that we'd seen in La Covaciella, when we visited Las Arenas de Cabrales.

I noticed it was the most elegant of the four we'd seen, the most complex and also, as a result, the simplest. The palaeontologist explained the origin of the image before inviting the children to reproduce it on a sheet of paper they'd been provided with.

"In prehistory," he said to them, "there were people who knew how to tell stories, others who were good at hunting, others who knew how to make fire, et cetera. And there were those who were good at painting, like the man who painted this bison. We're going to see," he added, "if you're good prehistoric painters. You've got five minutes."

The teacher had dimmed the classroom lights slightly so that the lines of the bison would stand out more sharply against the whiteboard backdrop. From one corner of the room, Arsuaga and I watched the intensity with which the class examined the model to try to reproduce it faithfully on their page. Close to me, there was a girl who had the tip of her tongue sticking out, and she moved it from side to side in her mouth in time with the movement of her pencil. The group's concentration was total. Arsuaga had told them it was a competition without a prize. Or that the prize consisted of doing it well, which seemed to spur them on more than if he had offered them a real trophy.

One by one, they began to hand in their work. The teacher brought the classroom lights back up, and we looked each of them over in turn.

They were a disaster.

The palaeontologist smiled as though he had just confirmed a hypothesis.

"It's very hot in here," he said, taking off his grey round-neck jumper, beneath which he had a dark-blue shirt covered in a pattern of small leaves. "I take it you like my shirt," he said.

"It's great," I admitted.

"It's very botanical."

Next, we carried out the same experiment with a prehistoric deer, twenty thousand years old.

"Now," said the palaeontologist, "imagine we're back in prehistory, and we're all part of a tribe. We live in caves, whose walls we like to decorate, and we're going to do an exercise now to choose the person in the tribe who's best at drawing."

While the pupils had their eyes trained on the whiteboard, and then their sheet of paper, and back again, Arsuaga explained it to me: "Traditionally, cave art was thought to have evolved toward complexity, toward perfection of a realist kind. They called that 'stylistic evolution'. As you can see, this deer is fairly simplistic: it isn't very detailed, there's little relief, the sections of the body aren't that clearly distinguished. It's just a single outline. Neither hooves, nor eyes, nor ears are very well defined. A lot of people would say: 'Anybody could do that, even a nine-year-old child.' What do you think?"

"I don't know. Its simplicity is too good."

"Indeed. Stylistic simplicity doesn't imply mental simplicity. This figure seems very elemental, doesn't it? In principle, it'd be a cinch to reproduce it."

"Yes, in principle."

"Well, the five minutes are up — let's go and collect their drawings."

We gathered them up, and confirmed that not one of them had captured the spirit of the original. There hadn't been a single child capable of reproducing the elegance of the prehistoric deer.

Still, Arsuaga insisted on a third drawing — this time of a bear from the French cave at Chauvet.

"I brought it," he said to me, "because it seems easier than the previous one, though it would be beyond me to execute it. It's extremely complex in its simplicity."

"It's fantastic," I agreed.

"Amazing, huh?" exclaimed the palaeontologist. "Know how old it is?"

"How old?"

"Thirty-one thousand years."

"And?"

"Either the dating is wrong, or the idea of evolving toward realism is wrong. It's thirty-one thousand years old, and just as perfect as the pictures from fourteen thousand years ago."

When we gathered up the work, the result was identical to the previous exercises.

The teacher sent the children off to recess, and they went enthusiastically, leaving us alone.

"All this," said Arsuaga, "is even more amazing if you think that prehistoric artists didn't use grids to capture the proportions, which means that to reach this level of perfection they must have practised a lot."

"Where?"

"We don't know — maybe in the sand on the beach, or on the banks of rivers, using a stick. They must have practised somewhere, because in order to paint a bear like the one we've just been looking at, you'd need to have drawn it lots of times before. You can't do that off the cuff."

"And?"

"And it's a mistake to associate a child with a prehistoric being."

We put on our coats and went out into the playground, where sleet was falling. The kids were chasing a ball from one side of the yard to the other. Arsuaga called one of them over and asked him to stand next to him. Then he looked at me.

"If you look," he says, "I'm practically twice as tall."

"Yes."

"But their brains are already 95 per cent the size of mine. Their capacity for mathematics is identical to mine. In the space of a month, they'd be able to do things of an astonishing complexity. They'd know how to add a half and a third, for example. Could you do that?"

"I don't think so."

"Doesn't there seem something out of kilter to you in the fact that these children have the brain of an adult in the body of a child? This is one of the mysteries of developmental biology."

"Right."

We went to the kitchen to get a coffee with the other teachers, as it seemed to be their breaktime. They asked us what we were doing there, and I told them about the experiments Arsuaga had carried out with the three- and nine-year-olds. On the way, I told them about the size of their pupils' brains.

"In all other mammals," added the palaeontologist, "this development is gradual. The brain and the body grow at the same rate. At the time of what we call 'the growth spurt', which coincides with puberty and which is a characteristic of the human species, the child's brain is already the size of an adult one. It's a strategy in our development: we have to socialise. And the smaller the body, the better, because it costs less: fewer calories are consumed. Children up to the age of eleven or twelve stay very small, in such a way that they don't participate in the social game and they don't constitute a threat to the adults. But then, in two years, they change — and how they change! Before that age, children often don't want to eat, or they're really bad eaters, much to their parents' despair.

"Whereas teenagers, if you aren't careful, they'll empty your whole fridge. In our species, their bodies double in size over the course of two years. It's a hell of a thing. Honestly, it's incredible we even survive puberty. It's a crisis — there's no other way to put it — and some writers compare it to the metamorphosis of insects. Because of this, from a pedagogical point of view, to educate children as if they were adults would be as ridiculous as educating a caterpillar as if it were a butterfly. A caterpillar isn't a butterfly in miniature; it's something else altogether. Nor is a child a human being in miniature; it's something else. Ortega, quite rightly, was opposed to the idea of forcing children to read *Don Quixote*, because it's an adult book. When I hear long-suffering mothers complaining that their adolescent children are acting like cocoons, I say: 'Don't worry, señora, out of that cocoon a beautiful butterfly will surely emerge.'"

Back in the car, headed homeward, I shivered with cold.

"You're cold because you're not wearing a Timberland," smiled the palaeontologist, showing me the label on his anorak.

FIFTEEN

The miracle diet

"I saw an oyster thistle when I was on campus," said the palaeontologist melancholically.

"Oyster thistle?" I said. "Sounds familiar."

"It's the only thistle that has yellow flowers. In Castilla, they just call it the little thistle, and around this time of year, end of February, coming into March, they put it in salads. You don't see that anymore, because it was a poor man's meal."

We were in La Gran Tasca, a Madrid restaurant on the Calle Santa Engracia, very close to the busy Plaza de Cuatro Caminos. Arsuaga said he'd called up the day before to pre-order the speciality of the house: a stew of which we had just been served the soup, with which we prepared our stomachs, and into which I, being hungry, dipped a bit of my bread. We clinked glasses with a Bierzo wine that tasted of liquorice.

"You were talking about that thistle, the ... what was it called?" I asked.

"The oyster thistle."

"The oyster thistle. You were referring to it with some nostalgia."

"It's just incredible how things come back around. I first

heard speak of it in 1970, from a professor at my university. I'm a *rara avis* because everyone says they learned nothing in university, whereas I learned everything there."

"And it hasn't turned out too badly for you," I pointed out.

"That's what I mean. Everything I know about biology I learned there. I've built on all that study since then, obviously, but fundamentally ..."

At that moment, the waitress arrived with a vast oval clay dish that she set down in the middle of the table and whose contents we stared at for a few moments, stunned.

"Amazing!" I exclaimed, in response to this extraordinary spectacle, part meat and part vegetable, a realist vision that nonetheless reminded me of the images in the paintings of Arcimboldo.

The palaeontologist smiled.

"Make a note," he said. "Chickpeas, potatoes, cabbage, blood sausage, chicken, beef, chorizo, bone marrow, pancetta, salt pork, and what we call 'la bola' in Madrid, which is a combination of spiced meat and bread."

"It's also got pepper in it."

"Not *also*," he corrected. "The pepper's a key component. What you've got before you here is a Neolithic meal, even though the pepper came from the Americas — a stew comprising vegetables cultivated by human beings and cuts from domesticated animals. This one we've been served here is special, of course, in sheer abundance and variety. The Palaeolithic diet, on the other hand, was based on game and vegetables they gathered. It was, in other words, an extractive economy: they took what they needed from nature. So now, who would eat a dish of recently picked chickpeas?"

"Nobody. They needed fire to cook them," I deduced.

"We'll take fire as a given," said the palaeontologist. "But in order to put them over the fire you need a receptacle, a container. A stew pot seems like the most normal thing in the world to us now, but its appearance implies not a biological but a technological-cultural revolution, and a hugely significant one. Most of what was cultivated in the Neolithic could not be eaten raw."

Meanwhile, we had attacked the stew, each according to his individual taste. I had poured out a ladle of chickpeas along with salt pork and cabbage, to soften and add flavour to the legumes. The palaeontologist, who was more meticulous, took small portions of whatever there was in the earthenware dish, arranging them round his plate according to rather mysterious criteria. He started with the beef, then he added the chicken, the chickpeas, the chorizo, the pancetta, the pepper ...

"You eat things in alphabetical order?" I joked.

"I like seeing all the separate ingredients before mixing it up. Just to get an idea."

And after a few seconds of Buddhist observation, he did indeed mix up the whole lot, and started to eat with the expression of someone enjoying a metaphysical thought.

"It's magnificent," he finally exclaimed.

"Yes, it's very good," I agreed, being more eager, and having already devoured half a plate and about to go for a refill.

"All the nutrients are in the ground," Arsuaga said. "Plants take the nutrients from the ground, from the minerals in the ground, from the water ... Thanks to the energy-giving properties of sunlight, they convert inorganic material into organic material."

"Photosynthesis," I said, remembering a lesson from school.

"Photosynthesis. All plants, whether cultivated or not, are the same. When the ground is fertile, productivity is high. If it's not deep enough or the soil is poor …"

"It's a bad business," I said, using the end of my knife to extract the marrow from a bone like somebody scooping butter out of a pot.

"Yes, it's no good at all. What do we do in the Neolithic?" he continued. "We modify the economy of nature. An area of land that produces a great quantity of plants, a forest, with its different strata capable of feeding a multitude of species, is transformed into something that will feed one species only."

"Chickpeas, for example."

"For example."

"That means," I added, euphoric at the grub and the wine, "that nature tends toward polycultures, and we tend toward monocultures."

"Put it however you want," said Arsuaga. "The point is that this is a way of ensuring great yields for humans, but for humans only. The vegetation in a forest feeds an incredible variety of vertebrates and invertebrates. But cereals and legumes will only feed you and me."

"Disaster!" I offered in a tone of feigned lamentation.

"It's a brutal economic transformation. An ecosystem is an economic system; even if we call it an ecology and it sounds more refined, it's still a question of economics. Of resources."

"So?"

"So, we fell a forest and turn it into fields we can cultivate. Where previously there were thousands of plants and animals, now there's only one plant and one animal. The biomass is the same, but this biomass is edible only for us. We've taken all that forest's resources for ourselves."

Arsuaga ate slowly, because he never stopped talking; but he was very efficient, because he was better at selecting and dissecting. As I watched him, I felt sorry not to have gone slower myself, as I was practically full. To compensate, my hands were free to take notes.

"But," I asked, "is that a good thing or a bad thing?"

"Is what a good or a bad thing?" he said, cutting open some blood sausage to release the flavours inside.

"The thing about taking all the resources from the forest."

"It's brilliant, and it has certain antecedents: those of the economy based on small items — do you remember the small items?"

"Yes, snails, insects, bulbs …"

"Right, so now you go and tell a hunter that he has to feed himself on chickpeas. How many is he going to need to equal the same calories a deer would give him?"

"Tonnes."

"But chickpeas have the advantage of being storable, and they can last a long time, like all legumes."

"And disadvantages?"

"They can't be eaten without cooking, and in order to cook them, as we were saying, you don't just need fire but also a stew pot."

"In other words, you've got to invent ceramics," I concluded.

"Precisely. You get the appearance of baked-mud containers that also serve for storage. And with the storage of things, along comes the concept of good."

"The concept of a surplus," I said, more precisely.

"But there's somebody who *owns* the accumulated surplus, and so, with a surplus, you get social stratification, hierarchy."

"Are there no Neolithic societies where a surplus belongs to the collective?"

"Not many. Anthropologists distinguish between the

various stages of social evolution. There was the nomadic band, a mobile group that led a life not dissimilar to that of Palaeolithic hunter-gatherers. Then the clan appeared, which held all property in common, and which comprised a collective of individuals related through marriage that considered themselves descendants of some mythical figure. Then the tribe, which spanned various localities. Later, over and above the tribe, a chieftainship appears. These chiefs have a lot in common with what the Romans, when they arrived on the Iberian Peninsula, would call 'kinglets'."

"You're not having your marrow?" I asked, in case I might be able to have it myself.

"I'm going to. Give me a second."

Suddenly, thanks to the stew, I was discovering that the palaeontologist had a Zen side to him. There was a part of him that was constantly meditating, even when he was talking or eating. That explained a lot about his personality — a certain distance, a certain ironic tone, a certain devotion — which until now had been a mystery to me. I started looking at him differently, a bit like the sage from that iconic TV series *Kung fu*.

How strange everything is, I thought.

"Bone marrow," he continued, "is the Palaeolithic part of this stew. We like it because ingesting it entails a huge calorie hit. Everything else is Neolithic — that is, it comes from domesticated animals."

"But you can get marrow from a domesticated cow, too."

"Even so, the concept is Palaeolithic."

"And are you going to have yours, or aren't you?"

"I don't think so. I'm done, I give in," he said, placing his cutlery on either side of his plate.

"I'll have it, then."

"OK, but make a note of this summary: the non-monetary economy, which produces surpluses, leads to the creation of the state via the succession of institutions we have just touched on: clans, nomadic bands, chieftainships, and finally kingdoms or republics, either way. The state, ultimately."

"And all that because we've discovered that vegetables and cereals can be cultivated and stored."

"Up until this point, there was only one way — a very interesting way — of stockpiling a surplus. Let's see, what would you do with the leftovers of an elephant you had just killed, after you and your fellow hunters had eaten your fill?"

"I'd smoke the meat."

"Methods for smoking meat still haven't been invented. What would you do?"

"I don't know. Deposit it in the bank?" I joked.

"And what was the bank in which elephants were deposited in those days?"

"No idea."

"Simple: you'd call in another tribe. They would eat it, and then they'd be in your debt."

"The bank was the stomach of the people in the neighbouring tribe."

"That's how a surplus was conserved in the Palaeolithic. This implies the appearance of a kind of accounting: my neighbouring tribe owes me one deer."

"That's not bad," I said. "But is it possible that the capitalist concept of interest might have appeared already, and that they'd have to pay back a deer and a half?"

"That, I don't know. What I do know is that storing something that you can't eat in someone else's stomach is a quite brilliant idea."

"I don't much like the fact that the side-effect of the invention of surpluses should have been the appearance of private property," I said.

"Plus you get the emergence of silos, granaries, of course," added Arsuaga.

"And that's where everything starts to go to shit. That's what Yuval Noah Harari says in *Sapiens*, and what Christopher Ryan hammers home in *Civilised to Death* — that the Neolithic saw the start of embourgeoisement …"

"The truth is, we see nothing, no biological indicator, in the remains of Neolithic people to suggest there was any improvement in their quality of life. They're smaller than the hunter-gatherers, their brains aren't as big, and they're riddled with rheumatisms because of the work they have to do as part of an agricultural system — milling the grain, looking after the livestock, et cetera. Plus, their life expectancy is no better than their Palaeolithic forebears."

"So why did the Neolithic triumph, then?"

"Because an area that's cultivated, or turned into pasture for livestock, will feed more human beings than a natural ecosystem. They didn't live better lives, but they could have more children, for example, and more continuously. But you've made me digress."

I kept quiet.

"Oh, I remember: I wanted to point out a problem with legumes, apart from the fact they have to be cooked in order to be eaten."

"Which is?"

"They don't taste of anything. These chickpeas we've just eaten are pure starch. To make them flavoursome you have to combine them with the pork, chorizo, blood sausage, and all the rest, because starch has the capacity to absorb lipids."

The waitress came over and asked if we're going to want dessert, and we replied as one with a yes (from Arsuaga) and a no (from me).

"You're still hungry?" I asked, gesturing that there was still more than half a dish untouched.

"Just 'cause I fancy it," he said.

"Well, bring two little spoons, then," I asked the waitress, "and put these leftover chickpeas in a couple of Tupperware boxes for us."

When the woman left, the palaeontologist, who never does anything without good reason, told me that ordering the fried-milk pudding had been no accident.

"Nothing's by chance," he added. "It's an excuse to explain that you and I aren't normal."

I thought for a moment that he was going to confess that he, too, was a closet Neanderthal. But instead what he said was that we are mutants.

"Mutants?" I asked. "You and me?"

"Look, in all mammal species the babies drink breastmilk. But only for a period."

"Right."

"Breastmilk has proteins and fats, as well as a glycide called lactose. In order to metabolise lactose, one needs a certain enzyme, a protein called lactase. Lactase is produced during lactation, and ceases to be produced upon weaning, at which point we mammals become lactose intolerant. Meaning that if you ingest milk as an adult, there's one thing that will definitely happen, and another that will probably happen. The definite is that you won't assimilate it, because it can't be metabolised without that enzyme. The probable thing is that it will irritate your digestive tract."

"I've heard that phrase a lot: lactose intolerance."

"It isn't intolerance; it's the norm. It's entirely normal not to be able to metabolise milk if you aren't feeding from your mother's breast."

I looked a little warily at the bowl of fried milk we'd just been brought.

"But you and I can," I said, in what was something between a statement and a question.

"You and I can because we're mutants. The culture has changed our biology. In central Europe, a genetic mutation appeared that meant people, a lot of people, continued to produce lactase all their lives. And it turned out very well for them: they produced a lot of offspring because they were cattle-rearing peoples, meaning they had milk in abundance."

"Which meant they had their proteins guaranteed."

"Of course. Milk contains proteins, fat, and glucose — everything, in other words. It's the most complete foodstuff going."

"I know people with a lactose intolerance," I said, recalling a niece of mine.

"That's completely normal. Those people haven't mutated. In central Europe and Scandinavia, they're all mutants. Then, the further you go away from that epicentre, toward Turkey and the Mediterranean, the percentage of mutants reduces, but in Spain it's still the majority. Most humans are lactose intolerant. You don't get milk in Chinese cuisine, for example, nor in the Americas. In India, it depends on what caste you are. Those that are Indo-European …"

"So in the production of lactase," I interrupted him, "there's a gene involved."

"Nonetheless, there are some people who have got to the

same point via different mutations, because the genome is a system — that is, there isn't a single gene involved in the production of this enzyme. The Masai, for example, spend all their time with their cows, to such an extent that they say God gave them the cows and that they are the owners of all the cows in the world. Well, the Masai also have a mutation that allows them to ingest milk after they've been weaned, but it's a *different* mutation from ours. A Masai smells of milk. They drink it mixed with blood that they drain from the cow's jugular with a little tube. They smell of fermented milk, like a baby that spends the entire day drinking milk and regurgitating it."

On leaving the restaurant, because the weather was fine, we decided to take a stroll before getting onto the Metro.

"Tonight I'll have those leftover chickpeas fried," I said, holding up my Tupperware box.

"I'm keeping mine for tomorrow — I'm stuffed. But we should have talked about the brain," Arsuaga said with frustration.

"It's just I spend the whole time interrupting you."

"That's true."

"I'd rather you told me a bit about hunger."

"Talk about hunger after that stew? You're crazy."

"It's just that if we'd talked about hunger before the stew, we would have been dissolved in our own gastric juices."

"Hunger," conceded the palaeontologist, "is behind everything. It's been a big problem for humankind."

"Are there any species who haven't experienced hunger?"

"No. In the northern hemisphere, the majority of living beings die of an illness that's called winter. It's what life is all about: making it through to the spring as best you can, at whatever cost. And only the few make it, very few. Spring is very generous, and autumn is rich in fruits. Summer can start to drag if August goes on longer than it's supposed to. But there's always the autumn, just ahead, and autumn is a time of abundance. Everything falls from the sky. Acorns, for example, are eaten in great quantities in Castilla, and in *Don Quixote*. They are sweet, plus you can use them as fodder for the pigs."

"Spring is such a joy," I said, relishing the late-February sun that heralds its arrival. We passed young people who were already wandering around with no coats on, some of them in colourful T-shirts.

"Springtime is good for carnivores, because they can eat defenceless newborn creatures," continued Arsuaga. "And for the herbivores, there's grass."

"And when there's grass for the herbivores, there's meat for the carnivores, right?"

"Exactly," he said. "Look, the movement of livestock from winter to summer pastures is a good example of the changes that occur over the course of the yearly cycle. The mountain pastures are good for the cattle during the month of August — hence the name *agostaderos*. But the grass gradually becomes scarcer, and the cows have a bad time of it at the end of summer. Autumn brings rain and all the fruits. Joy! Then comes winter, and that's just about surviving in any way you can. The oldest and the youngest die. Winter is also a real bastard in terms of all the snowfall, because the snow covers what little grass there is. Write this down: winter is the worst illness of all."

I made a note. Then I said: "Hence the importance of surpluses."

"But in the Palaeolithic, there weren't any surpluses, and

you couldn't just call Telepizza to come and deliver to your cave. You had to venture out for the things you needed — meaning you'd either get soaking wet or frozen to the bone."

"No Telepizza in the Palaeolithic," I also wrote.

"Let's do a very simple sum," he said. "Let's say you need three thousand calories a day because you're a woodcutter."

"I wouldn't be a woodcutter."

"OK, two thousand five hundred, then. And the brain consumes 20 per cent of that."

We'd stopped at a traffic light. Next to us was an elderly couple. The woman exclaimed: "Evaristo, that poor man. Life really is nothing at all, isn't it?"

"Twenty per cent is a lot for its weight, isn't it?" I said.

"A hell of a lot. Put your brain on some scales, if you can, then weigh the rest of your body separately, and you'll see."

"And that imbalance between weight and energy consumption, does it have anything to do with how much we happen to overthink things? Does a person with obsessive ideas consume more calories than a normal person?"

"No, the brain consumes the same amount independent of whether you think a lot or little. It consumes glucose, and lots of it. The neurons are insatiable, whether you use them or not. Now, imagine that an Australopithecus wants to evolve to become a Sapiens, for which it needs a bigger brain. But it needs to save energy from somewhere in order to achieve that. Where do you think he'll make the saving, bearing in mind that the amount of calories doesn't change — two thousand five hundred — and that they need to be worked for?"

"It's really so hard to get hold of them?"

"Not nowadays; it's easy as anything nowadays. But put yourself in the Palaeolithic. Blood, sweat, and tears."

"So, the solution," I concluded, "does not include increasing the number of calories, because there's nowhere to get them."

"Right. The saving happens in the digestive tract. We can imagine there are three chapters in the economy of the human body. The first corresponds with the vital organs: liver, kidneys, heart. You can't skimp there; they're all completely non-negotiable. The second chapter, the brain, we want to give *more* to that, not less, so that it can expand. So what's left?"

"The digestive tract."

"Exactly, and this is what we did: made the digestive tract shorter. If you take the digestive tract of a lion, from the oesophagus to the anus, stretch it out flat and measure it, and then you do the same with that of a zebra, you'll see that the zebra's is much longer than that of the lion, because it's a herbivore, so it has to metabolise huge quantities of the grasses and fibres that it lives off, all the cellulose it consumes. A zebra needs a long digestive tract, because what it eats isn't very calorific. There has to be a choice between what is abundant and not very calorific, and what is scarce and very calorie-rich. Such is life."

"And is that what we did?"

"That's what we did. Change the diet, which was herbivorous to begin with, for a different, higher-quality one, and as a consequence make the digestive tract shorter."

"And did this saving translate into the brain getting bigger?"

"Correct. Which in turn led to a more active social life, and gave rise to politics."

"And this came before cooking with fire?"

"At the point when we started cooking with fire, the brain had already got bigger."

"I thought the growth in the brain was a consequence of eating cooked foods?"

"No, it got bigger when we started eating energy-rich foods, cooked or not. You can eat meat raw, like lions do, with their short digestive tracts. All carnivores have short digestive tracts."

"If I tell you the length of a digestive system, could you tell me what the animal it belongs to would eat?"

"Sure, give me one."

"I can't think of one right now."

"Write this down: carnivores don't need to cook. Wolves don't cook. But it's true that cooked food digests better; this is a fact. On this particular matter, some people, myself included, think that fire is very ancient and that it's found at the origin of cerebral expansion, while others are of the opinion that fire came into feeding *after* the brain had expanded. That doesn't make it any less important. We are children of fire."

"And when this growth in the brain happened, did we immediately become Sapiens?"

"No, we were still hominids. Pre-Sapiens, if you like. This is three hundred thousand years ago we're talking about here."

"But three hundred thousand years ago, symbolic thought had already appeared."

"Somewhat," said the palaeontologist after a pause and a gesture of hesitation, as if wondering whether or not he ought to tackle such a subject at this time of the afternoon.

Neither of us said anything on the walk to the Metro. As we started down the stairs, he returned to the subject of fire to explain that it allows us to soften our food and absorb it better.

"This is clear," he added. "But be sure to make a note about this debate between those who think fire made *Homo sapiens*, and those who claim fire only appeared on the last furlong of evolution."

"It's still just a question of nuance," I said.

"But nuance is everything."

"So we did talk about the brain after all."

"Of course — what did you think? I never leave a class half-taught."

Then we parted ways, because we were headed in different directions. When I arrived on my platform, I saw him on the opposite one. We exchanged a smile, and each held up our own Tupperware container of Neolithic chickpeas, as if toasting with them.

His train arrived first.

SIXTEEN

Passing into posterity

The inscription on the gravestone read: "Luisito Meana González. Havana 31/12/1926–Madrid 9/1/1936. Your parents have not forgotten you."

"Just a kid," I said, pointing at the photo of the boy. "Poor thing."

"At least he got to miss the civil war," said the palaeontologist.

We were walking from grave to grave, searching for an epitaph that we could make our own, but every one of them featured somebody who was not being forgotten by somebody else.

"One of the most usual kinds of immortality consists of living on in the memory of others," said Arsuaga. "Hence this formula you see so often, 'They won't forget you.' Your parents won't forget you."

"But it's a homely sort of immortality," I said, "a domestic immortality. And realistic, no doubt. Nothing to do with the posterity that writers from other ages aspired to. I think that until well into the twentieth century, most novelists were producing their work for posterity, and there are some who believe in it still."

"For example?"

"I don't know," I hesitated. "Maybe Vargas Llosa. But posterity is dead. Now we're living in post-posterity. The 'they don't forget you' does nonetheless continue to prevail, because it's an aspiration to something possible. So long as you aren't forgotten by your parents or your children, you'll be OK."

We were in Madrid's Almudena cemetery, where Arsuaga had brought me in his Nissan Juke. It was empty because it was eleven in the morning on an unremarkable weekday in early March, and on unremarkable weekdays in early March, at this time, people have other things to be doing. Nevertheless, strange situations did occur: not long before, we'd seen a woman rushing between the graves, carrying a shopping bag with a baguette sticking out. Since this cemetery is, in terms of its size and layout, a real city, we wondered, naturally, if there was a branch of Carrefour someplace where the deceased could, through sheer force of habit, sustain the customs they'd had when alive.

It was cold and sunny.

I asked Arsuaga what we were doing here, and he said the cemetery was a good place to write the last chapter of our book.

"Everything ends with death, right?" he concluded.

"I don't know if everything does," I said, "but stretching my legs always helps. I do find the calm in these places relaxing."

"Plus, there's something very interesting I'm going to tell you about."

"Which is?"

"All in good time. Now, let me have a look at the map; there's one grave in particular I'm looking for."

While he consulted the map, he kept talking: "In Spanish, we use a similar expression for 'I've bought a lot at the cemetery' and 'I've secured tenure'. The latter is something we state employees say quite often, and we usually follow it with 'Because I'm in post already.'"

"So it's only a short hop from being in post to being in the cemetery," I said.

The palaeontologist, who is a university professor, gave a look to suggest he agreed, but that he wasn't giving in. Ever since I'd discovered his Buddhist side, I'd interpreted his way of expressing himself differently.

In spite of the map, we were lost, as if in a labyrinth. After various circuits, we found ourselves once again standing at Luisito's grave.

"We're never going to get out of here," I said.

"It's what tends to happen to those who come into this place," said Arsuaga, pointing to an area in disrepair with open, empty niches looking at us like dark eyes. "Those empty niches must be the ones bought by people in perpetuity. I think I'm right in saying that perpetuity, in a cemetery, means ninety years. Then, if nobody lays claim to them, they take the remains to a common grave, and sell the niche again."

"Even that obsolescent perpetuity," I said, "already lasts longer than posterity."

"Seems to me you're the one who's obsessed with posterity," he pointed out.

A cemetery car went past at that moment, driven by one of its employees. He stopped, because he saw how clueless we appeared.

"Looking for something?"

Arsuaga showed him a point on the map, and the man invited us to get in his car.

"I'll take you," he said, "because you're going to get yourselves lost — this place is vast."

When we were all in the car, he explained how the sections of the cemetery were arranged, divided into "districts" and "zones". I asked him whether there was a lot of tourism in the Almudena, and he replied yes, that sometimes they came in groups, with guides.

"They wanted to bring in one of those open-air double-decker buses, you know, like the ones you see in the centre of Madrid, but I don't know what happened."

After crossing several different "districts", we finally arrived at the grave the palaeontologist had been looking for, which was that of none other than Ramón y Cajal. We got out of the car and stood respectfully in front of it. On the stone, among other names and dates, we read: "Santiago Ramón y Cajal, 1852–1934".

"Totally neglected," said the palaeontologist sadly.

"Are you guys family?" the employee asked.

"Something like that," I replied, faced with the self-absorbed silence of Arsuaga, who was checking the foundation of the grave and testing out the whole thing's solidity.

Finally, he asked: "How much would it cost to get this all cleaned up?"

"You'd need to get rid of the damp ... I don't know," the employee started to calculate. "The front is made with a kind of brick that gets damp and buckles in the rain ... I'd say a thousand Euro. If you wanted to restore the lettering, which is badly worn, twelve hundred altogether."

"Look," said Arsuaga, while taking photos of the grave with his phone, "there are ants all over the base."

The employee withdrew respectfully, after having handed us a card.

"Ramón y Cajal," the palaeontologist said, in exasperation,

"who that man didn't know at all, and who's generally considered a Nobel laureate like any other, is actually one of the great geniuses of all time. He was on a par with Newton, Einstein, Darwin … Among the five or six most important figures in history. He's the most cited author in all scientific journals across the world, much more than Newton."

I remembered that our subject was posterity, but said nothing because I could see that the palaeontologist was a little upset. So we remained standing in front of the grave in a silence that he was the one finally to break.

"I read somewhere that the Academy of Sciences had decried the state of the grave, and demanded that the government do something about it," he said. "For the love of God, it's only twelve hundred Euro! I thought they were talking about millions! The academy could just pay for it. It make me feel like paying for it myself. Is that a laughable amount or what?"

"Yes," I said, "it's laughable."

"Can you imagine Newton's grave being left like this? They've got it in Westminster Abbey. Darwin's, too."

"That's what we are like, Spain and I, señora!" I quoted the famous Marquina line.

"Ramón y Cajal's whole scientific career was a struggle between two contrary theories to explain the way the brain functions: the neuron theory and the reticular theory. He argued for, and ultimately proved, the neuron theory."

At that moment, incomprehensibly, a young man passed by in shorts and a T-shirt, jogging. The palaeontologist and I exchanged an incredulous look.

"Well," I said, "that must be the deceased son of the dead woman who was coming back with her shopping."

The comment, which was supposed to be funny, had no impact whatsoever on Arsuaga, who was still hooked on his discontent.

"I went to great lengths" he continued, "to get the government to buy the house where Ramón y Cajal died. It's on Alfonso XII, near Atocha. His whole world was around Atocha. He had his laboratory there and, a stone's throw away, the university. His children inherited the house, and it was passed down to his grandchildren, and then finally put up for sale. I managed to get in there one day to see the state it's been left in."

"And what happened?"

"Nothing. I did what I could, and then some. I even took it as far as the minister ... I also proposed it to a chain of private hospitals, and to the ministry of health. I know they were made aware of the proposal, but they barely even considered it. Some Mexican property developer bought it, and they've converted it into luxury flats. You can see it on the Idealista website. I think there's one of those commemorative plaques on the outside of the building, 'Santiago Ramón y Cajal lived and died here, blah, blah, blah'."

"Which year was that?"

"Twenty eighteen. Those total dicks. And you really would have to be a big swinging dick in society if you wanted to buy Newton's or Darwin's house. But Ramón y Cajal's house, who gives a shit! And his grave, which you'd need only twelve hundred Euro to clean up, well, you can see how it is."

"Yeah, really sad."

The palaeontologist took my arm, and we started to walk away from the genius's grave.

"Coming here," he said, "at least brings us onto the subject

I wanted to explain to you."

"Tell me."

"Ramón y Cajal has a lovely book called *The World as Seen by an Eighty-Year-Old*. In it, he talks about what it feels like to be old. Old age and death are two of the big problems for science. Why do we get old and why do we die?"

"Well," I said, "they're saying now that ageing is reversible. A lot of researchers are talking about it like a curable illness."

"Exactly. But how is it that every species has its own particular old age? Why does a rabbit live to five, and a human live to ninety? Where's the clock? Is there some design? If old age were an illness, can it be passed on? Would it be contagious?"

"Oh, I don't know …" I said.

"People say stuff just for the sake of talking," said Arsuaga. "Now we're going to talk seriously."

"Right," I agreed.

"Why do we have to die? Why don't the cells of the organs repair themselves in order that we can avoid death, in the same way they repair themselves after a wound? Here, let's sit down on a bench."

We sat down on a stone bench that was a little cold for my liking, but I decided not to complain.

"You tell me."

"Tell you what?"

"Arsuaga, have you noticed you get these absences?"

"Absences how?"

"Absences of a Buddhist nature, that's what I'd call them. Suddenly you escape as if you were going into one of those trances in transcendental meditation. I admire that in you."

"Don't talk nonsense. What were you wanting me to say?"

"The reason for ageing and death."

"Nobody knows. They are the two big enigmas of science. The two greatest enigmas of biology since Darwin."

"And I thought I was in for some great revelation."

"Look," he said, glancing around absurdly, like someone who was about to confess a secret, "since we're going to get rich with this book, in the next one let's do some research. Let's go on a trip around the world, go to all the best places, ask all the right questions, and we'll then publish the most exhaustive study ever about old age and death."

"We should start walking," I said, as I was getting cold.

We got up. The palaeontologist continued to speak in a low voice, as if the dead could hear us.

"We shouldn't be giving away these key points yet, because they're going to make up the core of our next work."

"But it wouldn't be bad for us to make a bit of a start."

"I feel like it would be a shame to spoil what could be absolute dynamite when the time comes. We'd need to travel a long way, make lots of enquiries, because it's a subject that touches on so many different things."

"You think immortality is enough for a book?"

"It's enough for a library," he laughed. "Whether we can talk about it in a way that will appeal to people is another matter, but I think we can. But, joking aside, let's move onto the subject that brought us here."

"I still don't know what that is."

"Lifespan and life expectancy."

"Right," I said, disappointed.

At that moment, an empty bus, the number 110, went past.

"I didn't know buses came through the cemetery," said Arsuaga.

"Nor me."

"What number was it?"

"The 110."

"Might there be a number 666?"

I laughed. We laughed.

"Natural selection," continues Arsuaga, "consists of the survival of the fittest. We've had 4 billion years for selecting the fittest. How is it possible, then, that we're still totally fucking useless? How is it possible that a virus could take us out? Why do we only last ninety years? What's really happening here?"

"That's what I want to know: what's happening?"

"We'll get there. First, though, what we were talking about, which was …"

"Looking at the difference between life expectancy and lifespan."

"Very good. If we're able to bring some clarity to this particular subject, which has generated so much confusion, I'll be satisfied."

"Doesn't lifespan depend on an increase in life expectancy?"

"No, that's a misconception. Lifespan — make a note of this — is something particular to the species. Different species have different lifespans. A dog lasts about fifteen years; a cat, slightly more; an elephant, seventy-odd, which is about the same as whales and dolphins. That's roughly how it is."

"According to that, the lifespan of our species has never changed? It was the same three hundred or three thousand years ago as it is now?"

"It really was. It was the same, although life expectancy in 1900, for example, was about thirty years."

"And how do you explain that contradiction?"

"I spend my life trying to get my students to understand

this: the explanation lies in infant mortality. What we call the life expectancy of a particular population is in reality the average age of death for its individuals. If you have very high infant mortality at a certain time, it brings that average down. And vice versa."

"In other words, lifespan in the Stone Age, for the human species, was the same as it is now; it was just that a lot of children died."

"Spot on."

"That wasn't so hard."

"Well, I'm sure you'll forget it sooner or later, and you'll be out there parroting the idea about our generation living longer than our parents' did."

"Well, that's what people think."

"People think wrong. Don't forget: life expectancy is the number of years, statistically speaking, you've got left to live, and that isn't the same if you work it out when you're one and when you're sixty. It's constantly changing. Infant mortality is a devastating thing in all mammal species, and ours is no exception. In the Palaeolithic, people didn't live to thirty, as is so often claimed; rather, it was that infant mortality was very high, and that skewed the average."

"So a guy from Altamira wasn't already an old man at thirty?"

"What are you saying? He would have been in far better shape than a fifty-year-old today. He spent his life exercising, he ate lean meat, he lived out in the fresh air, there was no pollution. He wouldn't have had an ounce of fat on him. Modern-day medicine recommends living just like Palaeolithic man."

"Well, I'd always heard they got old earlier."

"That's precisely what I'm trying to disabuse you of, but no doubt when you get home you'll be calling me up to get me to explain it all over again."

"But the general perception," I insisted, "is that each generation lives longer than that of their parents."

"Because people know nothing about statistics, which is a marvellous science," said Arsuaga. "Statistics is poetry; it's music."

"That's not what the students are always saying. It's a much-loathed course."

"That's because they don't get how it all works. Your mind needs to have the right sort of disposition. The fundamental *doh-ray-me* of it goes right over people's heads. Let's say you tell me an Ethiopian man is coming to see us."

"OK. There's an Ethiopian man coming to see us."

"So, then I, without laying eyes on him, tell you that he's going to be between x and y feet tall, and I'm right 95 per cent of the time. I can know how tall an Ethiopian is who I've never met! Doesn't that strike you as marvellous?"

"If you put it like that …" I said, uncertainly.

"And it's all thanks to statistics, which is what insurance companies make a pretty good living from."

"So the theory that each generation lives a year longer than its parents' generation …?"

"Don't give me that — I've just explained it. For our next book, we'll go and visit a car factory. Haven't you noticed that a moment always comes with cars when everything stops working? If it isn't one thing, then it's another?"

"Yes, except for your Nissan Juke."

"At the Ford factory, when the Model T was about to come out — the first-ever car produced for mass consumption —

Henry Ford went and asked his engineers which part was the least long-lasting. So they showed him one, whatever it was — I have no idea about cars. 'How long will this part last?' he asked them. 'Four years,' they said. 'Very good,' he said, 'I want all the other parts to last that amount of time, too.' In other words, he wasn't prepared to make parts that would last a hundred years, if they were going to end up on the scrapheap after four."

"And does something like that happen with the human body?"

"Clearly. We could make organs that last longer artificially — we've got the technology. But why would I want to go on living if my brain doesn't work? Nature knows what it's about. It has economised, like Ford."

"I see," I said.

"We'll carry out research in all the car factories and all the gyms of the world, because I for one want to know why I'm going to die."

"Maybe for you, since you're younger than me, the matter will be resolved. Genetics are very advanced."

"But not regeneration. The regenerative processes aren't so advanced," said Arsuaga.

"What do you mean?" I resisted. "I did a report on that worm, the *elegans*, whose ageing process is very like ours, and it turns out they've been able to extend its life by a huge amount, just like with the fruit fly. They live longer and better than the others of their species."

"In the lab. We don't know what would happen to them out in nature. There's a very high price to pay for living longer; it doesn't come for free, because an organism is always an integrated whole. Living longer without paying a price isn't possible biologically. All these little animals they've engineered to live longer wouldn't last five minutes out in nature."

"But we human beings, in a sense, do live in a lab."

"In a lab, maybe, but not in a hospital, with tubes coming out of every orifice. A lab rat is a disgrace to rats. I don't want to be a disgrace to humans."

"OK," I said, "I give in."

"Immortality has been promised us, in different ways, for as long as the world has been the world," insists the palaeontologist. "What's the difference between the person who tells you you're going to live for one hundred and twenty years with no price to pay, and the one who promises you a paradise in which you're surrounded by a hundred beautiful virgins? What's the difference? The only thing that's for certain is that both prophets belong to the same category of shameless reprobate. I want to make sure this is really clear, because we're in the final chapter of the book, and I want the ending to be good."

"Don't worry, I'll be faithful to your words."

"OK. Well, let's leave it there — I'm late for a meeting."

We walked together to the cemetery exit, where we said goodbye with a hug that was unusual, because the palaeontologist always keeps his distance.

Back home, I called him up: "Hey, Arsuaga, I'm going back over my notes, and I don't completely understand the difference between lifespan and life expectancy."

The palaeontologist gave a snort that didn't sound in the least bit Buddhist to me.

"I'm kidding," I said quickly.